The Institute of Biology's
Studies in Biology no. 65

Plant Tissue Culture

Dennis N. Butcher

Ph. D.

Principal Scientific Officer, Agricultural
Research Council's Unit of Developmental
Botany, Cambridge

David S. Ingram

Ph. D.

University Lecturer in Botany and Fellow
of Downing College, Cambridge

Edward Arnold

First published 1976
by Edward Arnold (Publishers) Limited,
25 Hill Street, London W1X 8LL

Boards edition ISBN: 0 7131 2558 6
Paper edition   ISBN: 0 7131 2559 4

Printed in Great Britain by
The Camelot Press Ltd, Southampton

General Preface to the Series

It is no longer possible for one textbook to cover the whole field of Biology and to remain sufficiently up to date. At the same time teachers and students at school, college or university need to keep abreast of recent trends and know where the most significant developments are taking place.

To meet the need for this progressive approach the Institute of Biology has for some years sponsored this series of booklets dealing with subjects specially selected by a panel of editors. The enthusiastic acceptance of the series by teachers and students at school, college and university shows the usefulness of the books in providing a clear and up-to-date coverage of topics, particularly in areas of research and changing views.

Among features of the series are the attention given to methods, the inclusion of a selected list of books for further reading and, wherever possible, suggestions for practical work.

Readers' comments will be welcomed by the authors or the Education Officer of the Institute.

1976   The Institute of Biology,
                                   41 Queens Gate, London, SW7 5HU

Preface

Plant tissue culture methods have advanced considerably in recent years and are now firmly established in the repertoire of biological techniques. Originally organs and tissues were cultured in order to study fundamental problems of plant morphogenesis. However, it is becoming increasingly clear that such cultures, grown under precisely controlled conditions and in the absence of contaminant micro-organisms, provide excellent experimental materials in many other aspects of plant biology.

We have presented the basic concepts of plant tissue culture with an emphasis on practical procedures. We suggest that problems associated with tissue cultures are very suitable for sixth form projects and similar research topics since they provide an excellent training in experimentation. With this in mind we have designed practical exercises which illustrate basic principles, and involve a minimum expenditure on apparatus.

We wish to thank Professor P. W. Brian, F.R.S., for encouragement, and Mr. A. Sogeke and Mrs. A. Hand for trying out the practical exercises. We also wish to thank Professor H. E. Street and Dr. P. King for reading the manuscript and making helpful suggestions.

Cambridge, 1974   D. N. B. and D. S. I.

Contents

1 Introduction—Why Culture Plant Cells?

The reasons for wishing to culture cells in isolation from the plant were first clearly stated by the German botanist Gottlieb Haberlandt in 1902. Although he was unsuccessful in his attempts to culture cells he foresaw that they could provide an elegant means of studying morphogenesis. The principal rationale for using cell cultures for this purpose is the belief that a knowledge of the properties of cells isolated from the plant will give important information regarding the properties and interrelationships of cells within the plant.

There was little progress towards culturing isolated cells or tissues until the early 1930s when technical developments introduced by a number of investigators led to the successful aseptic culture of isolated roots of several species. The next significant advances came in 1938 with the culture of wound callus of carrot and tumour tissues from a tobacco hybrid. These successes were followed by a period of development when culture media and technical methods were improved, permitting the culture of a wide range of organ and callus cultures from many different species. Since 1960 methods have progressively become more advanced and have culminated in highly specialized techniques for culturing single cells, cell suspensions and naked protoplasts.

During the last decade it has been realized that some of the special properties of plant tissue cultures make them highly suitable as experimental material for many fields of study in addition to morphogenesis. This has led to a complete change in the status of tissue culture methods. Whereas, ten years ago, tissue cultures were the prerogative of a few investigators primarily interested in growth and development, they are now used by workers in a wide range of disciplines including biochemists, geneticists, plant breeders and plant pathologists.

In the past the terms used in studies of plant tissue culture have been poorly defined and often confusing. For the most part we will use the terms as they have been defined by STREET (1973) in the volume entitled *Plant Tissue and Cell Culture*. The following types of aseptic culture of plants may be distinguished:

Organ cultures—these are isolated organs, including cultures derived from root tips, stem tips, leaf primordia, primordia or immature parts of flowers and immature fruits.

Embryo cultures—these are cultures of isolated mature or immature embryos.

Callus (or tissue) cultures—these are tissues arising from the disorganized proliferation of cells from segments (explants) of plant organs. Callus cultures are usually grown as a mass of cells on a solid medium.

Suspension cultures—these consist of isolated cells and very small cell aggregates remaining dispersed as they grow in agitated liquid media. Suspension cultures are sometimes called cell cultures on the grounds that they represent a lower level of organization than callus cultures.

Like intact plants, cultured cells, tissues and organs require for growth the elements N, P, Ca, Fe, Mg, Mn, Cu, Zn, B, Mo, S and K. These are usually added to culture media in the form of mineral salts. There are also requirements for oxygen and hydrogen in the form of water, and for oxygen as a gas. Unlike most intact plants growing in the light, plant cultures also need carbon to be supplied in an organic form, usually as a sugar. In addition amino acids, B vitamins and growth hormones or complex extracts such as coconut milk are often required. In other words, while the majority of intact plants have an autotrophic nutrition, plant cultures are heterotrophic.

In the first part of the book we describe the various kinds of plant tissue cultures and give an account of their respective contributions to studies of growth and development. In addition we consider the uses of tissue culture in other areas of plant science. In the second part we deal with practical aspects of growing plant tissue cultures in the hope that it will encourage beginners to try their hand.

2 Organs and Embryos

2.1 Introduction

An essential feature of organ and embryo cultures is that they retain their characteristic structures and continue to grow in a manner comparable to that of their intact counterparts. In this respect organ and embryo cultures are clearly distinguishable from callus and suspension cultures where the organization and development of the intact tissues are lost. Organ cultures provide excellent experimental material in that they allow the properties and functions of the individual organs to be studied in isolation. They have been particularly valuable in studies of the interdependence of organs for growth hormones and other growth factors.

2.2 Roots

Isolated roots have a significant place in the history of culture methods since they were the first aseptic cultures to be maintained for extended periods by serial transfer. The pioneers in this field were Kotte and Robbins who in 1922 reported a limited amount of growth of root tips from aseptically germinated wheat seedlings. However, it was not until 1934 that White succeeded in culturing isolated roots of tomato for indefinite periods in a liquid medium containing inorganic salts, sucrose and yeast extract.

Detailed instructions for initiating and maintaining excised root cultures of tomato are given in Chapter 8, but briefly the procedures used by White were as follows. Tomato seeds were surface sterilized and allowed to germinate in aseptic conditions. The radicle tips, 10 mm long, were then excised and transferred to flasks containing nutrient medium. These 'tip cultures' grew at about 10 mm per day and lateral roots developed after 4 days. After 7 days (Fig. 2–1) they were used to initiate

Fig. 2–1 Tracing of a 7-day-old isolated root of tomato.

new cultures by excising the tips of lateral roots and transferring them to fresh medium. The laterals continued to grow at the same rate as the original tip to provide several roots which, after 7 days, were used to initiate stock or experimental cultures. Thus root material derived from a single radicle could be multiplied and maintained in continuous culture. Such genetically uniform root cultures are referred to as a clone of isolated roots. When the procedures outlined above were carefully standardized, large numbers of uniform root tips were available to initiate experiments which quickly established that the requirement for yeast extract could be replaced by the addition of three B vitamins, namely thiamine, pyridoxine and nicotinic acid, and the amino acid glycine.

After the basic techniques had been worked out for tomato roots, attention was directed towards other species. The results were mixed. Species can be divided into three groups according to the way their roots respond in culture. There are those such as tomato, clover and *Datura* which have high growth rates and produce numerous vigorous laterals. These can be grown indefinitely in culture. There are others such as pea, flax and wheat which can be grown for prolonged periods in culture, but not indefinitely. Here failure is due either to a declining growth rate or to insufficient or weak lateral roots. Finally there are roots, particularly of woody species, which hardly grow at all. Presumably they require growth factors which so far have not been identified.

Studies of the nutrition of isolated roots have provided basic information regarding the dependence of roots on the shoot for growth factors. A survey of the effectiveness of different carbohydrate sources, for example, has revealed that sucrose is far better than the others tested. It has also been shown that roots of most species require the vitamin thiamine for prolonged growth. Pyridoxine and nicotinic acid are beneficial, but not essential. It has therefore been concluded that intact roots normally depend on the shoot for adequate supplies of vitamins. However, a word of caution is necessary since it is possible that the limited biosynthetic capacity of cultured roots results from the culture conditions rather than the physical isolation from the plant.

Further experiments have suggested that excised roots of tomato and several other species are self-sufficient for the hormones, auxins, gibberellins and cytokinins, since under optimal growing conditions the addition of these substances either has no effect or is inhibitory. This interpretation has been supported by the extraction of significant quantities of auxins, gibberellins and cytokinins from excised tomato roots. In cases such as groundsel, where auxins promote growth, the endogenous auxins are presumed to be produced in sub-optimal amounts.

Isolated roots have also been used for studying the factors which are responsible for the initiation and development of secondary vascular tissues. Excised roots of many species do not develop a vascular cambium and therefore do not form secondary vascular tissues. This indicates that materials which are translocated from the shoot are responsible for the initiation and development of a vascular cambium in the root. This hypothesis was tested by Torrey and co-workers in an elegant series of experiments in which isolated roots of pea and radish were first allowed to grow in a medium where they did not usually form secondary vascular tissues; then substances thought likely to stimulate development were fed into the root system via the basal cut end (Fig. 2–2). When the bases of pea

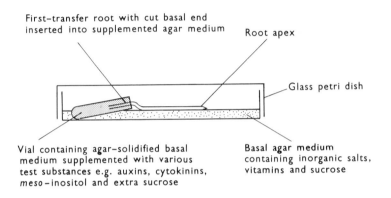

Fig. 2–2   A method used for initiating cambial activity in excised roots.

roots were fed with extra sucrose (8%) and IAA (10^{-5}M) a vascular cambium was initiated. In the case of radish roots, in addition to sucrose and IAA a cytokinin was essential and *meso*-inositol was beneficial. The treatment caused a twenty-fold increase in diameter and thus appeared to simulate the effects of the shoot system. These experiments have suggested that auxins, cytokinins and *meso*-inositol may have an important role in cambial development, although other factors are likely to be involved.

2.3   Leaves

Leaves may be detached from shoots, surface sterilized and placed on a medium solidified by agar where they will often remain in a healthy condition for long periods. Since leaves have a limited growth potential the amount of growth in culture depends very much on their stage of maturity at excision, young leaves having more growth potential than nearly mature ones.

Sussex and his co-workers have successfully dissected leaf primordia from the underground buds of the fern *Osmunda* and found that even extremely small primordia of about 1.2 mm are capable of developing normally on a simple medium containing inorganic salts and sucrose (Fig. 2–3). The only major difference between the intact and cultured

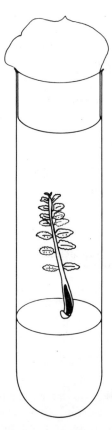

Fig. 2–3 Drawing of a cultured frond of *Osmunda cinnamomea* L. initiated from a young primordium. After 11 weeks on a medium containing inorganic salts and 2% sucrose a primordium would give rise to a frond 45 mm long with 10 pairs of pinnae.

leaves is that the growth of the latter is completed earlier, resulting in smaller leaves. Estimations of the numbers of cells in the intact and cultured leaves indicate that the differences in size are due to a reduced number of cells rather than a decrease in cell size.

Steeves has excised and cultured leaf primordia of different sizes from the stem apex of *Osmunda cinnamomea* in order to find out when the primordia become irreversibly determined. He found that the smallest primordia, which were flat mounds about 300 μm high, usually gave rise to shoots. However, with increasing size of primordia at excision there was an increased tendency to form leaves, and finally primordia excised when 800 μm long always gave rise to leaves. Steeves concluded that the primordia did not become irreversibly determined as leaf primordia until a relatively late stage in development.

Young leaf primordia of Angiosperms such as sunflower and tobacco have also been cultured. As with ferns the primordia develop into leaves having a normal morphology except that they are much reduced in size.

2.4   Shoot tips

The excised shoot tips of many plant species can be cultured on relatively simple nutrient media and will often form roots and develop into whole plants. Such shoot tip cultures have not been widely used in studies of nutrition and morphogenesis, although they do have a number of commercial applications.

For example Loo has found that stem tips, 5 mm long, can be excised from aseptically grown seedlings of *Asparagus officinalis* and cultured on a medium consisting of inorganic salts and 2% sucrose. To sustain growth indefinitely the cultures must be exposed to light. These tip cultures form cladophylls which may be subcultured at regular intervals; these often form roots and thus become plantlets, providing a convenient way of propagating valuable asparagus plants vegetatively.

Morel discovered another important application when attempting to culture stem apices of the tropical orchid *Cymbidium* (Fig. 2–4). He observed that when placed on an agar medium containing inorganic salts and glucose, the excised stem tips first proliferated to form callus, but there then followed a period where growth was localized in discrete areas over the whole surface. These protuberances eventually grew out to form the organized juvenile structures known as protocorms, which were excised individually and transferred to fresh medium where they developed into normal plants. Subsequently he found that the rate of propagation was improved by growing the tips in agitated liquid media. Under these conditions the proliferating callus masses repeatedly broke up to form a large number of individual protocorms. Morel's work has been particularly rewarding since it has been possible to adapt it for the commercial production of orchids, resulting in marked reductions in price.

The technique of vegetative propagation from stem tips is not restricted to asparagus and orchids, since it has been found that cultured apices of several other plants such as cauliflower, carnation and tobacco give rise to numerous plantlets which can be nurtured to maturity.

8

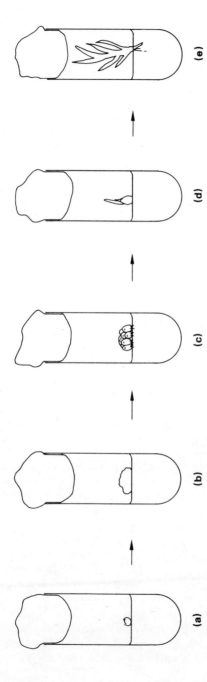

Fig. 2–4 Vegetative propagation of the orchid *Cymbidium*. (a) Apex excised and placed on a medium containing inorganic salts and glucose. (b) Proliferation of cells resulting in callus. (c) Development of a large number of protocorms bearing buds and rhizoids. (d) Protocorms separated and transplanted on to fresh medium. (e) A normal plantlet formed from a protocorm after 3 months in culture.

(a)   (b)   (c)   (d)   (e)

An important by-product of the culture of stem tips has resulted from the observation that the apices of some virus-infected plants become free of viruses when cultured on filter paper bridges standing in a liquid medium (Fig. 2–5). This has proved a very valuable method for obtaining virus-free stocks of plants where viruses have accumulated during prolonged periods of vegetative propagation, as in carnation, rhubarb, potatoes and geraniums. Its success arises partly from the fact that the apices of many virus-infected plants remain free of infection, although many other factors such as culture conditions may be involved. The effectiveness of the tissue culture procedures is enhanced when stem tips are taken from heat treated plants or when chemotherapeutants such as 2,4-D, malachite green or thiouracil are incorporated into culture media.

2.5   Complete flowers

Nitsch in 1951 reported the successful culture of the flowers of several dicotyledonous species. Not only did the flowers remain healthy in culture, but they developed normally to produce mature fruits. The procedure developed by Nitsch is as follows. Flowers, 2 days after pollination, are excised, sterilized by immersion in 5% calcium hypochlorite, washed with sterilized water and transferred to culture tubes containing an agar medium. The flowers of several species develop on media containing only inorganic salts and sucrose. Often the fruits which develop are smaller than their natural counterparts, but the size can be increased by supplementing the medium with an appropriate combination of growth hormones such as auxins, gibberellins and cytokinins. As might be expected, flowers put into culture before pollination do not usually produce fruits. Parthenocarpic fruit development has been observed, however, especially in the presence of synthetic auxins, which frequently induce parthenocarpy in intact plants.

The culture of young floral buds of cucumber has been employed to study sex determination. The young buds require a complex medium containing inorganic salts, B vitamins, tryptophan, casein hydrolysate and coconut milk. It has been shown that potentially male buds tend to develop ovaries in culture and that this tendency is enhanced by early excision or the addition of the auxin indole acetic acid (IAA). In contrast, gibberellic acid counteracts this trend towards femaleness. The development of potentially female or hermaphrodite buds seems to be more stable, since it is unaffected by the addition of IAA, gibberellic acid or other hormones.

2.6   Anthers and pollen

In order to culture anthers young flower buds are removed from the plant and surface sterilized. The anthers are then carefully excised and

10

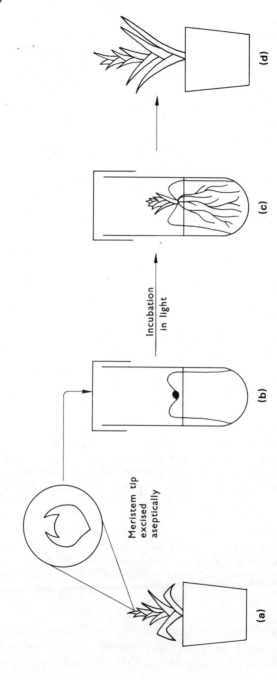

Fig. 2–5 Production of pathogen-free plants by the culture of meristem tips. (a) Stunted, virus-infected plant. (b) Meristem tip transferred to filter paper bridge in a tube of culture medium. (c) Rooted plantlet. (d) Rooted plantlet transferred to soil to produce a vigorous pathogen-free plant.

transferred to an appropriate nutrient medium. Those excised at an immature stage usually grow abnormally and there is no development of pollen grains from pollen mother cells.

Mature pollen grains (microspores) of several species of Gymnosperms can be induced to form callus by spreading them out on the surface of a suitable agar medium. Pollen of *Ginkgo biloba* proliferates on a medium containing inorganic salts, B vitamins, coconut milk and sucrose, while those of *Taxus brevifolia* and *Torreya nucifera* require a medium supplemented with the synthetic auxin 2,4-dichlorophenoxyacetic acid (2,4-D). The calluses consist of colourless parenchyma-like cells which can be subcultured indefinitely, but do not form roots, buds or embryo-like structures (see Chapter 3).

Mature pollen grains of Angiosperms do not usually form callus, although there are one or two exceptions. The pollen grains of *Brassica oleracea* may be induced to form small cell clusters if cultured by the hanging drop method. A drop of culture medium inoculated with pollen is put on a coverslip which is then inverted over the well of a cavity slide and sealed. Immature pollen of tomato can be induced to form callus colonies using a nurse culture technique. Anthers are placed on the surface of an agar medium and covered with a small disc of filter paper. A drop of pollen suspension containing about 10 grains is then pipetted on to the filter paper. Small colonies of green parenchyma-like cells develop within 28 days if the cultures are incubated in light at 25° C.

Recently Nitsch and Norreel have reported that pollen grains of tobacco and *Datura* can be induced to behave like zygotes instead of gametes. The pollen grains give rise to embryo-like structures (embryoids) which pass through the stages of embryogenesis to form plantlets and eventually flowering plants. They also found that a trauma such as a temperature shock given at the time of the first mitosis increases the number of cells following an embryogenic development.

Since anther cultures have been shown to have an abnormal development they have been of little value for studying the normal processes of microsporogenesis. However, anthers and pollen cultures have special features which make them very attractive for other purposes. Pollen grains cultured free or within cultured anthers give rise under certain conditions to haploid plants (plants having the same chromosome number as the gametes of a diploid plant). This discovery has aroused considerable interest among plant breeders. Its significance can be appreciated when it is realized that many cultivated plants do not breed true to type and inbreeding, even if possible, takes many years of patient work to produce a pure line. Sometimes this laborious procedure may be avoided by by-passing the sexual process and using haploid plants which possess only half the normal number of chromosomes. Haploid plants can arise naturally by the premature development of one or more cells in the embryo-sac, usually the egg cell. Obviously cells of haploid plants are unable to undergo meiosis and are therefore infertile, but treatment with chemical agents such as colchicine causes a doubling up of chromosomes resulting in diploid plants with two identical sets of chromosomes (i.e. the plants are homozygous). Unfortunately, the frequency of occurrence of haploids in nature is very low. Hence the anther and pollen culture techniques, which increase the number of haploids available to the breeder, are potentially very valuable.

Although these studies are still in their infancy, haploid plants have already been produced from pollens of several species, mainly through the independent efforts of Maheshwari, Nitsch and Nitsch and Sunderland. The list includes the following genera: *Nicotiana* (tobacco), *Datura*, *Atropa*, *Oryza* (rice) and *Lolium* (rye-grass). The procedure for obtaining haploid plants from cultured anthers of tobacco is briefly as follows (Fig. 2–6). Closed flower buds of an appropriate age are excised from the plant and surface sterilized by immersion in a solution containing 1% calcium hypochlorite. The sepals and petals are then removed and the anthers excised and immediately placed on to an agar medium. With anthers of tobacco development proceeds on a normal tissue culture medium lacking plant hormones. Anthers of other species have complex hormonal requirements, needing both auxins and cytokinins for development. The cultures give rise to large numbers of embryoids which go through stages comparable with those of zygotic embryos to form plantlets. The cultured anthers split open after 4 to 5 weeks at 25° C revealing embryoids at the cotyledonary stage. The plantlets may then be teased apart and transferred individually to fresh culture medium. When an adequate root system has developed the young plants are transplanted to compost and raised to maturity.

The actual stage of development of the anther at the time of excision is very critical if haploid plants are required. In tobacco this critical time seems to coincide with the first unequal division of the pollen grains. Surprisingly it is the normally quiescent vegetative cell which usually gives rise to the haploid embryoid, whereas the generative cell does not appear to participate.

In one experiment Sunderland raised 400 plants from tobacco pollen. Chromosome counts on the root tips revealed that the vast majority were haploids, a few were chimaeras having both haploid and diploid roots and flowers, and 1 to 2% were homozygous diploids. In other experiments it was shown that treatment with colchicine increases the amount of chromosome doubling if given during the early stages of embryoid development.

An alternative way of producing haploid plants is to culture the anthers in a medium containing auxins which induce the proliferation of haploid callus instead of embryoid formation. Once the callus has been formed it is transferred to a medium with no auxins and a reduced sugar level. Shoots, and later roots, are thus induced to develop and numerous haploid plantlets result. The callus approach is important in several respects. Firstly, it may be the only route possible for some species, such as rice. Secondly, callus cultures are less stable than organized tissues and the chromosome doubling frequency is high. Diploid homozygous plants may therefore be regenerated from 'pollen' callus without resorting to chemical treatments, which often introduce complications. Thirdly,

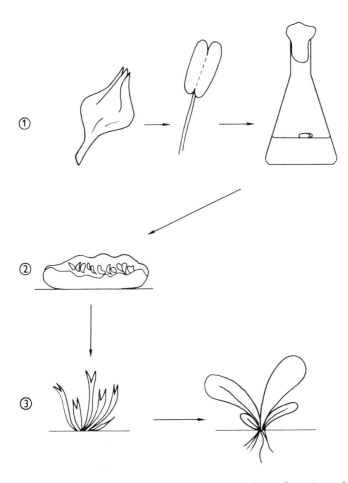

Fig. 2–6 Development of haploid plants from the anthers of *Nicotiana tabacum*. Stage 1: stamen excised from closed flower bud. The anther is then excised and placed on a nutrient medium. Stage 2: After 28 days the anther splits, revealing numerous embryoids. Each embryoid develops from a single pollen grain. Stage 3: after 35 days plantlets have developed from the embryoids and can be teased apart and transplanted individually to a medium lacking sucrose.

callus formation is an important step towards the establishment of haploid cell lines for use in mutation studies. Here, of course, the problem will be to maintain stability and to stop the chromosomes from doubling. The important problem of instability in callus cultures will be considered in Chapter 3.

If the investigations now being carried out with 'pollen' plants are successful they could ease some of the breeding programmes of genetically complex plants such as brassicas, sugar beet and cereals.

2.7   Ovules and embryos

A knowledge of the changes which take place during the development of a mature embryo from a single cell zygote is essential for an understanding of morphogenesis. During this critical period of the plant's life the genetic information contained within the zygote is programmed and expressed in a way which allows the co-ordinated unfolding of the complex events which result in a multicellular embryo possessing shoot and root primordia. Normally the developing embryo is embedded in the parental tissues and presents a difficult material for experimental studies. Clearly if ovules, or better still the young embryos, could be cultured *in vitro* they would provide elegant experimental systems. Maheshwari and his colleagues have successfully cultured ovules excised from *Papaver somniferum* (opium poppy) and other species and demonstrated all stages of development from fertilization to seed maturation.

Progress has also been made towards the culture of very young isolated embryos. Embryos have been dissected from the ovules of *Capsella* (Shepherd's purse) and put into culture (Fig. 2–7). It has been shown that the nutritional requirements of such embryos become progressively less fastidious as they mature. Very small globular embryos require a delicate balance of the hormones auxin, gibberellin and cytokinin, adenine together with B vitamins, and relatively high levels of sucrose. These requirements are successively lost as the embryo passes through heart and torpedo stages until the nearly mature embryo requires only carbon dioxide, oxygen, mineral salts and light to continue development. This indicates that during development there are progressive increases in the biosynthetic capacities within the tissues until an autotrophic existence is achieved in the young seedling.

In addition to its importance in developmental studies the culture of young embryos offers a method for raising hybrid embryos derived from certain inter-specific crosses which abort when left in the ovule, e.g., *Linum* (flax), *Hordeum* (barley).

The discovery a few years ago that large numbers of embryo-like

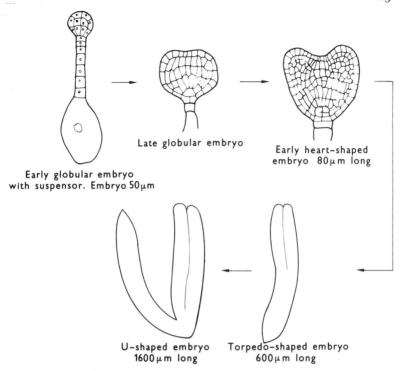

Early globular embryo
with suspensor. Embryo 50μm

Late globular embryo

Early heart-shaped
embryo 80μm long

U-shaped embryo
1600μm long

Torpedo-shaped embryo
600μm long

Fig. 2–7 Drawings of immature embryos of *Capsella bursa-pastoris* showing stages at which they have been cultured.

structures may be induced in certain callus and suspension cultures has provided a unique experimental system for studying embryogenesis. Studies with this system will be described in Chapter 3.

3 Callus Cultures

3.1 Historical

During the late 1930s Gautheret and Nobecourt in Europe and White in the U.S.A. pioneered techniques which revolutionized the methods used in plant tissue culture. Gautheret in 1939 reported that the tissues of carrot root explants proliferated to form large masses of disorganized tissue when placed on a nutrient medium containing mineral salts, glucose, cystein, thiamine and the auxin IAA. These proliferations, which became known as callus (wound callus), could be maintained in culture indefinitely by subculturing small pieces of tissue on to fresh medium at regular intervals. At about the same time White reported independently the cultivation of callus tissues from the pro-cambium of stem segments of the tumorous hybrid tobacco *Nicotiana glauca* x *N. langsdorffii*. In this case the tissues grew on the simple medium previously devised for isolated root cultures solidified with agar. Unlike the carrot callus these tissues did not require an auxin supplement. Subsequently callus isolates were obtained from a wide range of plants including Dicotyledons, Monocotyledons and Gymnosperms (Fig. 3–1).

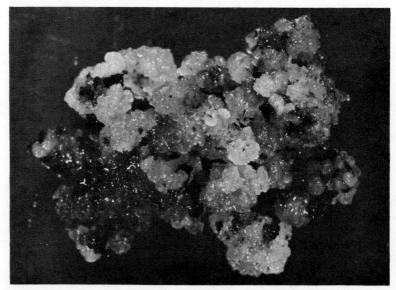

Fig. 3–1 A callus culture derived from a carrot root.

3.2   Initiation of cultures

Before attempting to initiate a callus culture it is first necessary to sterilize the plant organ from which an explant is to be taken. If taken from a seedling it is usually more convenient to sterilize the seed before imbibition and allow it to germinate in aseptic conditions. Then at a suitable stage the appropriate organ (cotyledon, hypocotyl or root) can be excised with a sharp scalpel and transferred to a nutrient medium solidified with agar. If, on the other hand, the explant is to be taken from a mature organ such as a carrot root, potato tuber or a woody stem segment, the whole organ or segment is surface sterilized before excising a piece of tissue from the inner undamaged and uncontaminated tissues (Fig. 8–5). Sterilizing agents commonly used are sodium hypochlorite (1.6% available chlorine), mercuric chloride solution (0.1% w/v) and a solution of bromine in water (1% w/v).

Depending on the explant, callus proliferation may arise from the cambium, cortex, pith, secondary phloem or even xylem parenchyma. The callus takes anything from 3 to 8 weeks to reach a size when it may be subcultured by transferring small pieces of tissue (50–100 mg) to fresh medium. Once a callus has become established it should be sub-divided and subcultured at regular intervals (e.g. every 4 weeks) until sufficient material for experiments becomes available. In general 25° C is a suitable temperature for incubating cultures and exposure to low light intensities is often beneficial, but not essential. In experiments growth is most conveniently assessed by measuring fresh and dry weight increases, although other parameters such as increases in cell number, cell volume and insoluble nitrogen have been used.

3.3   Culture media

Although the original callus cultures described by Gautheret and White grew indefinitely on relatively simple defined media, it was soon discovered that the nutritional requirements of many tissues are more demanding. In some cases callus growth occurs only when mixtures of amino acids, vitamins, kinetin or other growth factors are added to the medium. In others complex supplements such as coconut milk (liquid endosperm of the coconut), yeast extract or casein hydrolysate are necessary.

It is interesting to note that it was the search for the active factors in complex media supplements which led to the discovery of the important group of plant hormones known as the cytokinins. Miller and Skoog examined the active factors in yeast extract which stimulated cell division in explants of tobacco stem pith tissues in the presence of auxin. After several years of rather unrewarding work due to the small quantities of active material present in yeast extract, they turned by chance to some old commercial preparations of DNA. Surprisingly these old preparations, but not fresh ones, were found to contain relatively large amounts of substances active in the tobacco pith test. Subsequently they succeeded in isolating and identifying a highly active compound which they named kinetin (6-furfuryl amino purine), a derivative of adenine. So far as is known kinetin itself does not occur naturally, but a whole family of related cytokinins have been isolated from plant tissues; e.g. zeatin and N^6-(Δ^2-isopentenyl)adenosine.

The use of cytokinins and the development of better media containing higher levels of potassium, nitrate and phosphate have permitted the culture of many more calluses on defined substrates. The synthetic medium suggested for the practical exercises in Chapter 8 is a modification of that devised by Braun and Wood, but other media such as those of Murashige and Skoog or Nitsch are also effective.

An examination of the nutritional requirements of a large number of callus cultures derived from normal (i.e. not tumorous or habituated tissues, see section 3.8), reveals that an auxin and usually a cytokinin are required for maintaining high growth rates. The auxins most commonly used are IAA and the more stable synthetic compounds 1-naphthalene acetic acid (NAA) and 2,4-dichlorophenoxyacetic acid (2,4-D). The concentrations required depend both on the type of auxin and the origin of the callus, but usually a concentration within the range 0.01–10 mg per litre is suitable. The most frequently used cytokinin is kinetin, although benzyl adenine has also been used. Kinetin is added within the range 0.1–10 mg per litre.

Callus tissues either lack chloroplasts or have an insufficient number for growth in the absence of added carbohydrates. As with excised root cultures, sucrose is usually the preferred carbohydrate source, although glucose is sometimes equally effective. Thiamine, *meso*-inositol and other vitamins are often essential for maintaining high growth rates and are usually added to nutrient media. Other organic supplements such as asparagine, arginine, urea, glutamine, purines, pyrimidines and various amino acids have been beneficial in particular cases, but are not universally required.

Although calluses have been isolated from many different organs of a wide range of species it should be noted that many tissues remain difficult to culture. However, it is hoped that more cultures will be established as further information regarding nutritional requirements is accumulated.

In contrast to callus cultures from normal tissues, those from tumour tissues have very simple requirements, needing only inorganic salts, a sugar and one or two vitamins. They have no requirements for auxins or cytokinins. A more detailed account of the properties of tumour cells in culture will be given in section 3.8.

3·4   Morphology and cytology

Callus cultures can be defined as tissues which proliferate continuously in a disorganized fashion giving rise to an apparently amorphous mass with no regular form. Within this definition there falls a large number of morphological types. These vary according to external appearance, texture and cellular composition. Some calluses consist of hard compact tissues with small closely packed cells, while others consist of soft tissues with minimal cellular contact (Fig. 3–2). The pigmentation of callus

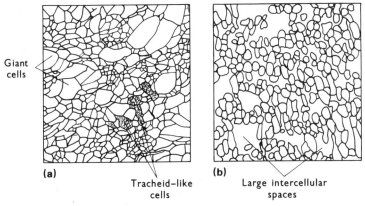

Fig. 3–2   Drawings of sections of compact callus tissue (a) and soft, friable callus tissue (b).

tissues is also variable, even among isolates from the same species. Many calluses lack pigmentation while others are pale green (chlorophyll), yellow (carotenoids or flavonoids) or purple (anthocyanins). The kind and degree of pigmentation is markedly influenced by nutritional and environmental factors such as exposure to light.

Microscopic studies have shown that callus tissues are often heterogeneous in cell composition. The diversity of the constituent cells varies according to many factors, including the origin and age of the cultures and the composition of the medium. Actively growing cultures usually contain a high proportion of vacuolated cells resembling parenchyma, and more localized groups of smaller dividing cells. The highly vacuolated cells have very diverse shapes ranging from spheres to filaments with all stages in between (Fig. 3–3). The dividing cells are either in the peripheral layers or distributed throughout the callus. Cultures sometimes contain tissues resembling xylem, phloem and cambium. The occurrence of such tissues is often influenced by the composition of the medium and age of the culture. For example high levels of auxin or a

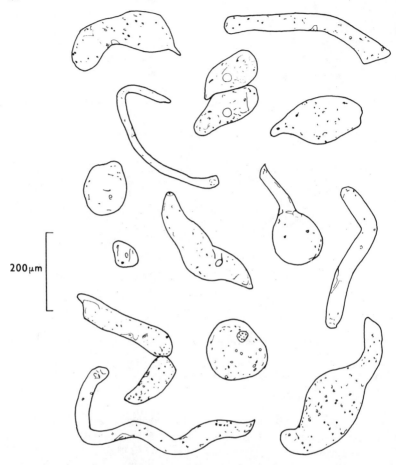

Fig. 3–3 Drawings of cells from a friable callus of carrot showing the diversity in shape and size.

prolonged culture period favour the development of groups of tracheid-like cells (Fig. 3–2). More will be said about the factors which influence vascular differentiation in the next section (3.5).

It is common for callus tissues to undergo quite marked changes when cultured for prolonged periods. Some of the more important changes occur in the nuclear cytology of the cells; polyploidy and decreases in the number of chromosomes, chromosomal breakages and rearrangements have all been observed. The frequency of such abnormalities usually increases with the age of the isolate. Although most callus isolates suffer from cytological instability, the rate at which the changes occur varies considerably from one species to another. For example, it is usual for callus cultures of carrot and tobacco to exhibit a significant level of polyploidy (three or more times the haploid number of chromosomes) within a few months of being isolated. On the other hand some callus isolates of *Crepis capillaris* and *Helianthus annuus* (sunflower) remain stable for periods of up to 2 years.

The chromosomal instability of cultured cells is likely to be a major obstacle to many of the potential uses of plant tissue cultures, e.g., in plant breeding, propagation and biochemical genetics, and it is essential that an understanding and eventual control of the factors involved is achieved. It should be noted, however, that Nickell and Heinz have exploited the genetic variations found in tissue cultures of sugar-cane in their crop improvement programmes. They isolated single cell clones (see Chapter 5, section 5.2) from a callus which had been initiated from internodal tissues. The clones differed from each other both morphologically and cytologically, as did the plants regenerated from them. Using such techniques it has been possible to obtain from a single sugar-cane variety a number of new lines having such attributes as increased disease resistance (see section 3.9).

3.5   Factors controlling cell differentation

Studies with callus tissues have led to a recognition of some of the factors which are important in determining the pattern of vascular differentiation. In 1949 Camus grafted small buds into the upper surface of callus masses from *Cichorium* (endive), composed only of parenchyma cells. After a period of incubation it was observed that vascular tissues were induced to form in the callus and were connected to the bud. Wetmore and his colleagues repeated these observations using buds and callus of *Syringa* (lilac) (Fig. 3–4a). Moreover, they found that they could replace the buds by inserting agar wedges containing sucrose and the synthetic auxin NAA. Further, by manipulating the amounts of auxin and sucrose, they were able to influence the type of vascular tissues formed. With a constant low sucrose level and increasing auxin concentration a ring of vascular nodules of increasing diameter was produced below the wedge. In contrast, with a constant level of auxin and increasing concentrations of sucrose, the vascular tissue changed from being exclusively xylem at 2% sucrose to almost entirely phloem at 4% sucrose. Sucrose at 3% gave a mixture of xylem and phloem. The most striking observation was that the regions of vascular tissue were arranged on one plane in a ring in such a way that the entire complex resembled normal stelar anatomy. This work has been successfully repeated with callus of *Phaseolus* by Jeffs and Northcote (Fig. 3–4b) with the additional

22

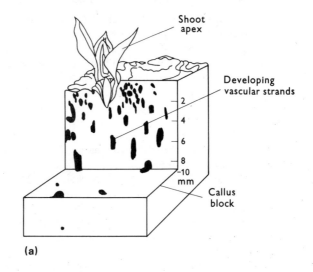

(a)

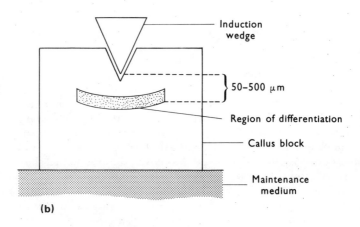

(b)

Fig. 3–4 Induction of vascularization in callus tissue. (a) Callus of *Syringa* into which has been grafted a stem apex bearing 2 or 3 leaf primordia. Drawing made 54 days after grafting. (After WETMORE, R. H. and SOROKIN, S. (1955), *Journal of the Arnold Arboretum*, **36**, 305.) (b) Diagram to show the induction of vascularization in a block of *Phaseolus* callus into which has been inserted an agar wedge containing auxin and sucrose. (After JEFFS, R. A. and NORTHCOTE, D. H. (1967), *Journal of Cell Science*, **2**, 77.)

observation that the sucrose effect is specific for α-glucosyl disaccharides. The hormone-like effects of sucrose are still not understood, although they have been observed in other experimental systems. Although the work described above suggests that auxins and sucrose determine the locations and type of vascular differentiation in callus it would be wrong to assume that only these compounds can influence such differentiation. Indeed there is evidence from other systems that cytokinins and other factors are involved. It should also be appreciated that these experiments concern differentiation of cells within relatively large masses of tissues and that we are still a long way from Haberlandt's original objective of being able to control the differentiation of single cells or even small groups of cells.

3.6   Organogenesis

In callus cultures cell divisions usually occur in a random fashion giving rise to disorganized masses of tissue with no obvious form or polarity. However, in certain experimental conditions shoot and root meristems or even embryo-like structures (embryoids) may be formed. Often these organized structures develop to form plantlets and eventually whole plants. Prior to 1955 the reports of organogenesis in callus cultures were sporadic, and little was known about the factors which influenced the initiation or controlled the development of organized meristems. However, this was all changed with the discovery of the cytokinins, which have dramatic effects on organogenesis. Thus Skoog and Miller have shown that a totally disorganized callus tissue derived from the stem pith of tobacco can be induced to form roots or buds, or simply continue to proliferate as a callus depending on the relative amounts of auxin and kinetin provided (Fig. 3–5). When the ratio of auxin to cytokinin is relatively high only root primordia are formed, and when the ratio of cytokinin to auxin is relatively high shoot buds are initiated. Intermediate ratios of the two hormones produce completely disorganized callus growth. The control of root and bud formation in callus culture by varying the auxin : cytokinin ratio has now been demonstrated for callus tissues of several plant species. The interacting effects of the hormones can be modified by other media constituents such as sugar, amino acids, casein hydrolysate and phosphate ions, although the auxin : cytokinin ratio is always the dominant factor. These findings, together with those described in the previous section (3.5), have led to the widely held concept that it is the ratio rather than the amounts of the different hormones which determines how plant tissues develop. Although there is good evidence for this view, a full understanding of the important interactions between hormones will not be possible while the primary action of the respective hormones remains obscure.

IAA concentration (mg/l)

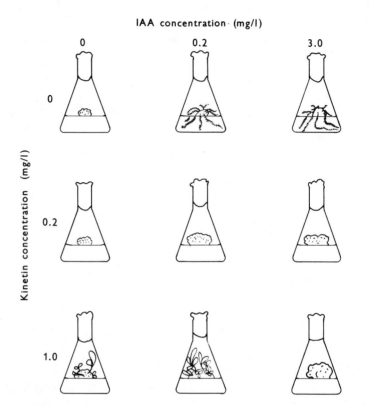

Fig. 3–5 Organogenesis in tobacco callus growing on media supplemented with differing concentrations of auxin and kinetin.

3.7 Embryogenesis

Reinert in 1959 discovered that undifferentiated callus cells of carrot become granular and show evidence of differentiation when transferred from a complex medium including coconut milk to a synthetic mixture containing amino acids, vitamins and hormones. A second transfer to another medium containing a low auxin level leads to the regeneration of a large number of plantlets. Microscopic examination of the cultures during the intermediary stages reveals the presence of bipolar embryo-like structures (embryoids). This discovery is important since it shows that the parenchyma-like cells of the callus can give rise directly to embryo-like structures, by-passing the sexual process. Steward and co-workers have shown that large numbers of embryoids can be induced to develop by plating out a suspension of cells obtained from carrot root pieces (explants) on a medium enriched with coconut milk. They have also shown that single cells from the suspension are capable of proliferating to form callus tissue within which embryoids can develop (Fig. 3–6). These embryoids, when transferred to fresh media, develop into mature plants capable of setting viable seed (Fig. 3–7). Recently Halperin has been able to induce embryoid formation without using coconut milk, by manipulating the levels of hormones and inorganic nitrogen in the culture medium.

Although embryogenesis has also been demonstrated in callus and suspension cultures of a number of other species, e.g., *Atropa belladonna, Ranunculus scleratus*, there are many instances where attempts to induce embryoids have completely failed, while some cultures which initially produce organs lose this capacity after a few months. In these cases the inability of the callus to form organs is ascribed either to metabolic deficiencies or genetic abnormalities. The essential difference between the two possibilities is that with the former it should be possible to induce organogenesis if we discover which metabolites are involved, whereas with the latter the ability to organize is irreversibly lost.

The fact that some cells within callus masses are capable of forming embryos or organized meristems is highly significant, since it indicates that such cells retain all the genetic information required for the normal development of the whole plant, i.e. they are totipotent. However, we know very little about the cytology or biochemistry of the potential organ forming cells, because the low percentage of cells which actively participate in organ or embryoid formation cannot yet be isolated or even identified. The earliest recognizable organ primordia consist of a small number of cells with dense cytoplasm which are distinguishable from their neighbours only on the basis of their staining properties. Methods will now have to be found for controlling organ initiation in small multicellular units, to allow the biochemical analysis of successive developmental stages. The large number of embryoids which may be induced in some suspension cultures such as carrot and *Atropa* is a step in this direction.

It is interesting that the early segmentation pattern of cultured embryoids of different species is usually very uniform. This contrasts with the situation within intact ovules where there is a great diversity of segmentation patterns during early zygote embryology among different species. This suggests that the characteristic segmentation pattern for a particular species may not be genetically determined but probably results from physical restrictions and polar chemical gradients imposed upon the embryo as it develops within the ovule; when these influences are removed, as in cell cultures, the embryology reverts to a common basic type of segmentation.

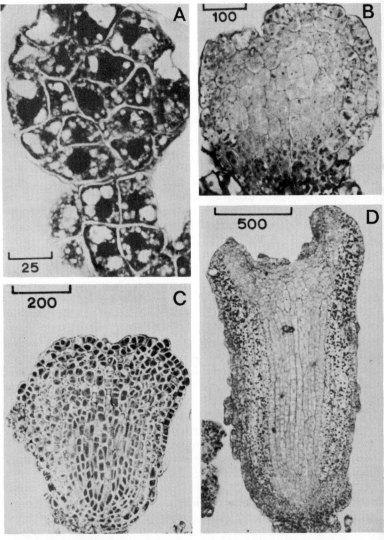

Fig. 3–6 Embryoids found in suspension cultures of carrot. **A.** Globular embryoid with 2 tiers of suspensor cells. **B.** Early heart-shaped embryoid. **C.** Late heart-shaped embryoid. **D.** Torpedo embryoid. Scales in μm. (Photographs kindly supplied by Professor H. E. Street.)

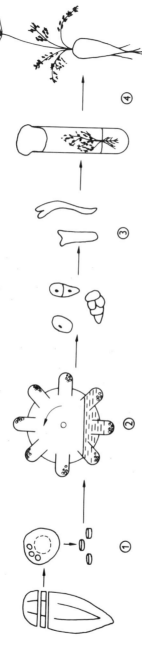

Fig. 3–7 Diagram to show how cultured cells from the secondary phloem of carrot can develop into embryoids, plantlets and mature carrot plants. Stage 1: surface sterilization of carrot and excision of thick transverse slice. Small discs cut from slice with cork borer. Stage 2: explants placed in rotating culture flask and incubated at 25° C. Stage 3: cells of explant proliferate liberating cells into medium to form suspensions. Embryoids develop within cell aggregates. Stage 4: embryoids are planted out onto agar medium where they become plantlets. These develop into normal plants which set seed.

3.8   Characteristics of tumour cells

As mentioned at the beginning of this chapter, one of the first tissues to be cultured as a callus came from hybrid tobacco plants which readily formed tumours. Since then the culture of tumour tissues has played an important role in the development of tissue culture techniques. One of the reasons for this is that they have simpler nutritional requirements than comparable normal tissues, and are therefore easier to handle in culture. Another equally important reason is that tumour tissues cultured in aseptic conditions provide an excellent experimental system for comparing the physiological and biochemical properties of tumour and normal cells. In this context we confine the term 'tumour' to excessive growths or proliferations of disorganized tissue which, once they have been induced, continue to grow in the absence of any inciting agent (bacterium, virus or chemical).

Three kinds of plant tumour have been studied using tissue culture techniques. These are crown-gall tumours caused by *Agrobacterium tumefaciens* (Fig. 3–8), wound tumours caused by the virus *Aurogenus magnivena* and genetic tumours which occur on certain interspecies hybrids, e.g., *Nicotiana glauca* × *N. langsdorffii*. These tumours, although caused in different ways, have many characteristics in common. Firstly, it is necessary for the host tissues to be mechanically damaged before they will develop; i.e. there is a wound conditioning requirement. Secondly, the tumour tissues, free of any causal agent, may be grafted onto healthy plants where they will continue to grow. And thirdly, when put into culture after elimination of the causal agent, tumour callus grows vigorously on a very simple defined medium containing neither auxins nor cytokinins. (Bacteria-free crown-gall tissues are obtained from primary tumours by heat treatment, by the addition of antibiotics, or by taking explants from parts of the tumour which have naturally become bacteria-free.)

The observation that tumour tissues have no requirements for auxins and cytokinins has led to the view that, unlike normal callus tissues, they synthesize unusually large amounts of these hormones and that this is responsible for their continuous proliferation. However, a full understanding of tumorigenesis requires an explanation not only for the disorganized growth, but also for the persistence of this property from one cell generation to another. The fact that the different kinds of tumour have many features in common has led to the view that the tumorigenic property is inherent in the genome of the host and that the various inciting agents (bacteria, virus or genetic instability) trigger off the tumorous condition. The observation that calluses derived from normal tissues occasionally give rise to sub-isolates which grow vigorously on media without auxin or cytokinins (habituated cultures) lends support to

Fig. 3–8   Crown-gall tumours on a sunflower stem.

this view. On the other hand there is another school of thought which considers that in the case of crown-gall tumours genetic information is passed from the inciting bacterium to the host cells during tumour induction. In this case it is the expression of this foreign genetic information which is thought to be responsible for the production of hormones and disorganized growth. The actual nature of the information passed from the bacterium to the host is still unknown, although nucleic acids, phages and plasmids have all been implicated.

3.9   Growth of pathogens in callus culture

Dual cultures, consisting of a plant parasite growing together with callus tissues of its host, can provide a simplified and controlled experimental system for studies of the interaction between the two organisms. The full potential of dual cultures for such studies has yet to be realized, although some important work is being carried out. For example, a group of scientists in the United States has recently shown that the genes controlling the resistance of tobacco to *Phytophthora parasitica*, cause of the black shank disease, are expressed normally in callus cultures of tobacco. It is clear that this work will now lead to a more rapid elucidation of the biochemical basis of the resistance process in tobacco than might otherwise have been possible. Some similar studies with other fungi and bacteria have met with only limited success because of abnormalities and variation within the cultured tissues. However, even such apparent shortcomings of tissue cultures are being exploited by breeders who are using plants regenerated from genetically abnormal callus cells as a source of new resistance to a number of pathogens. Notably it has recently been possible to obtain by this means clones of sugar-cane plants resistant to the virus causing the destructive Fiji disease. Application of mutagens to cultured tissues may result in many more successes of this type. Attempts to use callus cultures in studies of the interaction between plant viruses and their hosts have unfortunately met with only limited success, due mainly to a failure of the virus particles to replicate normally in cultured tissues. However, it is thought that isolated cells and protoplasts may prove useful as tools for the virologist, a possibility which is discussed in Chapter 6.

Callus tissue cultures have a special application in attempts to grow pure cultures of plant parasitic nematodes and fungal obligate parasites. The latter may be defined as fungi which are so specialized nutritionally that they can only grow and achieve full development in association with a living host. Thus the genetics and physiology of such pathogens cannot be studied in isolation from their hosts. The French botanist Georges Morel was the first person to establish a dual culture of an obligate parasite and its host when in 1944 he inoculated callus cultures of vine with zoospores of the downy mildew fungus *Plasmopara viticola*. Since that time many other dual cultures have been established and maintained (Fig. 3–9a). It has been the hope of almost all workers involved in growing such cultures that the fungus might grow out from the infected callus on to the surface of the culture medium, and there become established saprophytically. In most cases the best that has been achieved has been limited colonization of the medium by peninsulas of mycelium which have died upon being severed from the parent culture (Fig. 3–9b). However, the American plant pathologist Victor Cutter had rather more success than this during the

(a)

(b)

Fig. 3–9 (a) A callus of sugar beet infected with the downy mildew fungus *Peronospora farinosa*. (b) Growth of *P. farinosa* away from a sugar beet callus, on to the surface of the culture medium. (After INGRAM, D. S. and JOACHIM, IRENE (1971), *Journal of General Microbiology*, **69**, 211.)

1950s when, following Morel's advice, he established dual cultures from galls induced on Juniper by the rust fungus *Gymnosporangium juniperi-virginianae*. In seven of the many thousands of dual cultures grown on a variety of media *Gymnosporangium* grew out from the callus and became established as a saprophyte. Pure cultures of rust fungi are now proving very valuable in studies of host-parasite interaction. There is every reason to believe that dual cultures will be valuable in establishing similar saprophytic cultures of many other obligate parasites.

4   Cell Suspension Cultures

4.1   Introduction

The prospect of culturing plant cell suspensions in liquid media, employing the well-developed methods of microbiology, offers a unique experimental system for detailed studies of growth and differentiation. Ideally such suspensions should consist of single cells (rather than cell aggregates) which are physiologically and biochemically uniform. This potential use would be further enhanced if it were possible to synchronize the processes of division, enlargement and differentiation within the cell population. Although these ideals have yet to be achieved in full, sufficient progress has been made to suggest that they are more than a distant possibility.

In 1953 Muir demonstrated that cells of tobacco and *Tagetes erecta* can be cultured in the form of cell suspensions. Later similar suspensions were obtained from carrot root explants and callus tissues of *Picea glauca*, *Antirrhinum majus* and many other species. Subsequently it was shown by Nickel that the techniques primarily designed for the culture of micro-organisms could be used for growing relatively large quantities of cell suspensions.

4.2   Initiation of suspension cultures

Suspension cultures are usually initiated by transferring established callus tissues to a liquid medium, which is then agitated by one of the methods described below. In general it is the more friable callus tissues which give rise to suspension cultures with the greatest degree of cell dispersion. However, the dispersion of less friable tissues may be improved by modifying the culture medium. Raising the concentrations of auxin, altering the ratio of auxin to cytokinin or adding low concentrations of cell-wall degrading enzymes such as cellulase and pectinase are all effective. Nevertheless, it should be said that no suspension culture has as yet been shown to be composed entirely of free-floating single cells (Fig. 4–1). Even the most dispersed cultures so far established, such as those of sycamore, consist of cell aggregates as well as single cells. Clearly, the cells within the aggregates are in a different micro-environment from the free-floating cells. This undoubtedly contributes to the non-uniformity of cell size, shape and metabolism which is characteristic of cell suspensions. This heterogeneity is a

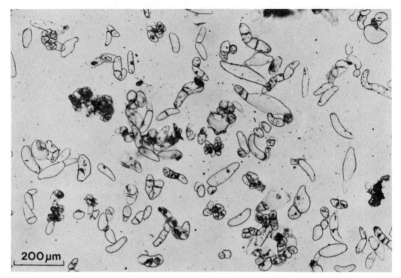

Fig. 4–1 Single cells and cell aggregates from a suspension culture of carrot.

formidable obstacle to the use of suspension cultures in precise studies of cell development. As with callus tissues, polyploidy and other chromosomal irregularities are commonly observed and contribute to the variability of cells.

In general, the media suitable for growing callus cultures for a particular species are also suitable for growing suspension cultures, providing that agar is omitted. However, in some cases suspensions are more exacting in their requirements; for example, the concentrations of auxins and cytokinins are often more critical.

Although the more dispersed suspension cultures consist entirely of thin-walled cells, others possess a proportion of lignified, tracheid-like elements. These usually arise in the larger cell aggregates. In certain cases the number of tracheid-like cells can be influenced by medium constituents such as auxins and cytokinins, but so far it has not been possible to induce a whole cell population to differentiate in a particular predetermined way. Thus it is not possible to control differentiation in cell suspensions with the precision required for detailed biochemical studies. Organs and embryoids have been induced in cell suspensions of some species, e.g., carrot and *Atropa belladonna*. Again these appear to arise in an unpredictable fashion from cell aggregates rather than from free cells.

Basically there are two types of suspension culture, batch cultures and continuous cultures. Both types are currently being used and each has its own particular advantages for specific research projects.

4.3   Batch cultures

Batch cultures are closed systems where the cell material grows in a fixed volume of medium which is agitated to maintain the even distribution of free cells and cell aggregates and to promote adequate gas exchange with the air. During incubation the amount of cell material increases for a limited period of time and reaches a point of maximum yield. At this point an exhaustion of a medium nutrient or some other factor, such as accumulation of toxic material, stops growth. If the culture is to be maintained, a small volume of the suspension must be removed and used to inoculate a flask of fresh medium. The growth pattern will then be repeated to yield a similar amount of material. In this way the culture can be continuously maintained and propagated by successive inoculations at appropriate intervals into fresh medium. Subcultures are made using a pipette or syringe with a canula orifice large enough to allow the free passage of single cells and small cell aggregates, but small enough to exclude the larger tissue masses and cell debris. Four different types of batch culture systems are currently in use, and these may be distinguished by the way in which movement of the liquid medium is achieved.

4.3.1   Slowly rotating cultures

Steward and Shantz, in 1956, described a specially designed flask, the nipple flask, for culturing free cells and cell aggregates released from large numbers of carrot explants (Fig. 4–2a and b). The flasks, of 250 ml capacity, possess eight nipples or projections and are mounted on large flat discs. When the discs are rotated at slow speed (1–2 revolutions per minute) the explants within the flasks are‚alternately bathed in culture medium and exposed to air.

4.3.2   Shake cultures

Muir described a much simpler, but equally effective, system for his tobacco suspensions. The suspensions, contained in Erlenmeyer conical flasks, are placed on an orbital platform shaker where they are agitated by a circular motion (Fig. 4–2c). Orbital shakers are widely used for cell suspensions of a large number of species. The optimum conditions for maximum growth and dispersion vary from one species to another, but a shaking speed of between 60–180 revolutions per minute with a 3 cm throw is suitable for most cultures.

4.3.3   Spinning cultures

Spinning cultures consist of relatively large culture vessels (10 litre bottles containing 4.5 litres of medium) fixed to a rigid frame. This is tilted at an angle of 45° and is capable of being rotated at 80–120

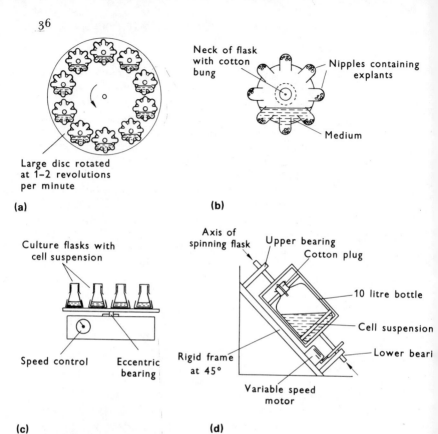

Fig. 4-2 (a) Large disc loaded with 10 nipple flasks used for growing cell suspension cultures of carrot. (b) Detail of a nipple flask. (c) Side view of a platform shaker loaded with suspension cultures contained in conical flasks. (d) Diagram of a 10 litre spinning culture apparatus.

revolutions per minute (Fig. 4-2d). The neck of the culture flask is sealed by a cotton plug which allows sufficient gaseous exchange for growth. The effectiveness of this apparatus is shown by the data reported for sycamore cells; in 21 days 4.5 litres of medium yielded 1.0×10^{10} cells having a packed cell volume of 1.4 litres and a dry weight of 40 g.

4.3.4 Stirred cultures

Large batch culture systems (1.5–10 litres) have been devised where the cells are kept dispersed and gaseous exchange effected either by bubbling air through the culture medium or by an internal magnetic stirrer. These cultures, because they are stationary, can be readily instrumented and connected to reservoirs of media and gas supplies. The temperature can be accurately controlled, either by an internal coil or by a water jacket. A flow diagram of a typical stirred batch culture is shown in Fig. 4–3a. The culture vessel is a 5 litre round bottom flask fitted with a large diameter flange joint and closed with a matching reaction flask lid with five ports. At the base there is a magnetic stirrer which revolves at 200–600 revolutions per minute. The air supply passes a vapour trap, a carbon filter and two microfilters to remove micro-organisms before entering the vessel via a sintered glass aerator. Air outlets are fitted with taps and plugged by non-absorbent cotton plugs. The whole system is made airtight so that the production of gases such as carbon dioxide or ethylene may be determined by analysis of the exit air stream. Culture sampling is effected by a temporary closure of the air outlet. This leads to a positive pressure in the vessel, causing the medium, plus cells, to flow into the sample receiver. The air outlet is then opened and the culture sample run off. The ports in the lid can be used to introduce a thermometer or the glass electrode of a pH meter. In more sophisticated designs automatic sample collection may be incorporated.

4.3.5   Growth patterns in batch cultures

The growth of sycamore cell suspensions in batch cultures has been widely studied. In non-synchronized batch cultures increases in cell number, dry weight, DNA, total protein and many other parameters follow the general pattern shown graphically in Fig. 4–4a. There is an initial lag phase, followed by a relatively short exponential phase of 3–4 cell generations, and finally growth declines into a stationary phase. Typically, the doubling time for sycamore cell suspensions during the exponential phase is about 40 hours. Data for suspensions of other species show similar patterns of development except that the doubling times vary considerably: *Nicotiana tabacum*—48 hours; *Rosa sp.*—36 hours; *Phaseolus vulgaris*—24 hours and *Haplopappus gracilis*—22 hours.

Although the general patterns of the changes for each growth parameter (cell number, dry weight, total protein, etc.) are similar for an individual species, the actual rate of increase during the exponential period can vary markedly from one parameter to another (Fig. 4–4b). It follows from this that the chemical composition of the cells changes throughout the growth cycle, indicating that the biosynthesis of at least some of the materials is not closely coupled to cell division. The observed uncoupling of biosynthesis and cell division may be modified by altering the composition of the medium. For example, with sycamore suspensions, doubling the initial concentration of 2,4-D from 4.5×10^{-6}M to 9.0×10^{-6} M increases the rate of cell division and reduces the cell doubling time by about 20 hours. However, there is no corresponding increase in dry weight accumulation and consequently the divergence between the rate of cell division and rate of dry weight accumulation increases. From this and similar studies it has been concluded that there

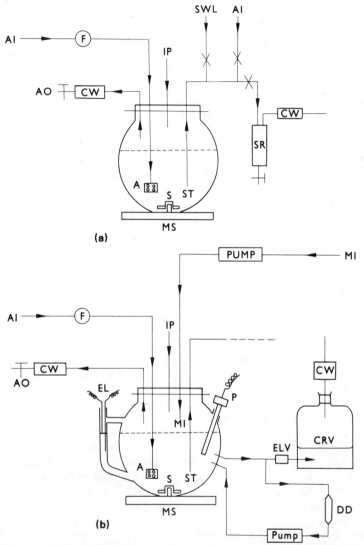

Fig. 4–3 Flow diagrams of advanced cell culture units. (a) Stirred batch culture. (b) Chemostat culture. Arrows indicate direction of flow of air or liquid; X indicates a clip on a silicone-rubber tube; ┼ indicates a glass tap. A = aerator; AI = air input; AO = air outlet; CRV = culture receiving vessel; CW = cotton-wool filter; DD = density detector; EL = volume-sensing electrodes; ELV = volume-controlling outlet valve; F = sterilizing glass-fibre air filter; IP = inoculation port; MI = medium input; MS = magnetic stirrer; P = probe, e.g. for oxygen tension or pH; S = stirrer magnet; SR = sample receiver; ST = sample tube; SWL = sterile water line. Note that connections to sample tube are not shown in (b). (Diagrams kindly supplied by Dr. P. King.)

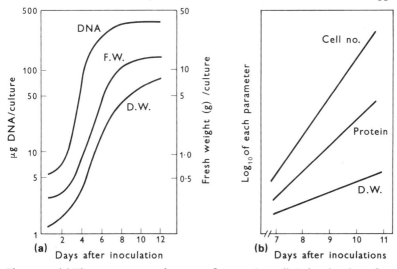

Fig. 4–4 (a) Time-course growth curves of suspension cells in batch culture for 3 parameters: DNA content, fresh weight and dry weight. (Adapted from FLETCHER, J. S. and BEEVERS, H. (1970), *Plant Physiology*, **45**, 765.) (b) Unbalanced growth of cells of *Acer* during the exponential growth phase of a batch culture. Semi-logarithmic plot showing rates of change of cell number, total protein and dry weight per unit volume of culture. (Adapted from KING, P. J., MANSFIELD, K. J. and STREET, H. E. (1973). Reproduced by permission of the National Research Council of Canada from the *Canadian Journal of Botany*, Volume **51**, 1973, pp. 1807–1823.)

are independent mechanisms for controlling cell division and many biosynthetic pathways.

Clearly the lack of homogeneity of cell populations and the continuous changes in physiological properties put severe limitations on the use of non-synchronized batch cultures for precise studies of biochemical aspects of differentiation. Recent work has, therefore, been directed towards devising systems where stages of cell division and cell enlargement are synchronized for prolonged periods. Several methods, including the temporary addition of cell division inhibitors and the temporary removal of cytokinins, have been used for this purpose, with varying degrees of success. Most progress has been made by Street and his co-workers; using a 4 litre batch culture unit fitted with an automatic sampling valve, they have been able to induce prolonged cell division synchrony in sycamore suspension cultures by a starvation and re-growth treatment (Fig. 4–5). This is a very significant advance in tissue culture technology. The objective now is to show a synchronization of events in the cell cycle other than cell division, i.e. mitosis, DNA synthesis and enzyme activities.

Preliminary work along these lines with synchronous sycamore

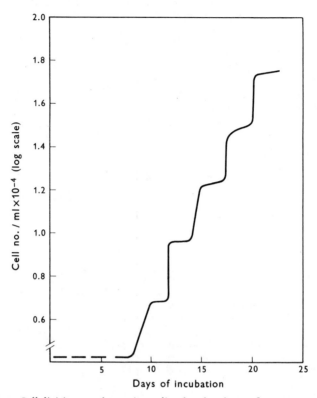

Fig. 4–5 Cell division synchrony in a 4 litre batch culture of sycamore. The cell numbers were estimated from samples taken by an automatic device. (Simplified from KING, P. J., MANSFIELD, K. J. and STREET, H. E. (1973). Reproduced by permission of the National Research Council of Canada from the *Canadian Journal of Botany*, Volume 51, 1973, pp. 1807–1823.)

cultures has shown that the synthesis of total extractable protein and RNA occurs throughout the interphase period of the cell cycle, but at increased rates during the later stages. Respiration rates also rise throughout the interphase period, but with two peaks of activity. On the other hand thymidine kinase and aspartic transcarbamylase show single peaks of activity. It is hoped that in the future investigation of the effects of hormones and nutritional factors on these metabolic events will give an insight into the mechanisms which regulate cell division and enlargement.

4.4 Continuous culture systems (chemostats and turbidostats)

A continuous culture system of the chemostat type consists of a culture vessel (Fig. 4–3b) into which a constant flow of nutrient medium is pumped, causing the displacement of an equal volume of culture (spent medium plus cells). The most significant property of such a culture is that its growth rate adjusts so that new cells are produced at a rate which compensates exactly for those being washed out. Thus a steady state culture is produced in which the density, growth rate, chemical composition and metabolic activity of the cells all remain constant. Steady state systems may also be established by linking medium flow directly to changes in a property of the culture such as optical density (as in a turbidostat) or medium pH. Automatic monitoring units adjust the medium flow in such a way as to maintain the optical density or pH at a chosen, pre-set level.

Street and his co-workers are now using chemostats to investigate the growth kinetics of sycamore cell suspensions. An advantage of chemostat cultures is that they may be used to stabilize the physiological states which have only a transient existence in batch cultures. This is achieved simply by fixing the dilution rate at the appropriate level. Preliminary studies have shown that the cells of different growth rates under steady state conditions differ significantly in their chemical composition and physiology. There is a decline in dry weight per cell as the growth rate increases and this is associated with a decline in cell volume at the highest growth rate. Respiration rate ($Q O_2$) and RNA values rise progressively with rises in growth rate. These physiological and cytological changes which occur when a population of cells is taken through a series of steady states are fully reversible. Furthermore studies of the transition from one steady state to another have shown that there are pronounced oscillations in the enzyme activity levels, characteristic for each enzyme monitored, which gradually decline in amplitude as the new steady state is established. Investigations of this type should eventually lead to a better understanding of the regulation of metabolic activity in higher plant cells, and may also provide information about the changes which occur during cell differentiation.

Thus it has been shown that the growth of higher plant cells can be manipulated by the application of techniques previously used only with micro-organisms. While the batch culture system seems to be the best method for synchronizing the activities of the cells within a cell suspension, the continuous system is more suitable for obtaining cell populations with particular required metabolic activities. The fact that the metabolism of the cells can be controlled by changing the input of certain medium constituents has significant implications for the use of plant tissue cultures in the production of commercially important products such as alkaloids, steroids and antibiotics.

5 The Growth of Single Cells

5.1 Introduction

One of the chief aims of Haberlandt in 1902 was to isolate single cells and maintain them in culture. This objective has been achieved only recently following the development of more effective nutrient media and specialized techniques for isolating cells from callus and suspension cultures. Three basic methods have been employed for culturing single cells, the paper raft nurse technique, the petri dish plating technique, and the growth of cells in micro-chambers.

5.2 The paper raft nurse technique (Muir 1953)

Single cells are carefully isolated from cell suspensions or friable callus with a needle or fine glass capilliary. Each is then placed on the upper surface of a square of filter paper which rests on an actively growing callus (nurse callus) (Fig. 5–1). This experimental system provides the single cell

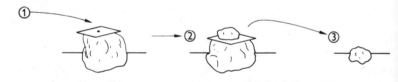

Fig. 5–1 Growth of single cells using a 'nurse' technique. Stage 1: a single cell taken from a friable callus is placed on upper surface of filter paper which is in contact with nurse callus. Stage 2: the single cell divides and daughter cells proliferate to form colony. Stage 3: when colony reaches a suitable size it is transferred to fresh medium where it gives rise to a single cell clone.

with growth factors produced by the callus as well as those from the culture medium. It was first used for crown-gall tumour cells where cells divided to give rise to small colonies. These colonies were subcultured on to fresh media to give callus isolates derived from single cells. A callus, originating from a single cell, together with its derivative cultures, is known as a single cell clone. Although the first successful work was done with crown-gall cells, the method has since been extended to cells derived from normal tissues of several species.

5.3   The petri dish plating technique (Bergmann 1968)

With this widely used method (Fig. 5–2) cell suspensions are first filtered so as to remove all tissue masses and large cell aggregates. This is important since the division of cells within aggregates is earlier and more frequent than the division of single cells. The single cells and aggregates of not more than six cells remaining in the filtrate are then incorporated into an agar medium which has previously been sterilized and then allowed to cool to 35° C. At this temperature agar remains liquid, but is not hot enough to kill cells. The mixture is dispensed into petri dishes to give a layer 1 mm thick and the dishes sealed with a plastic film which retards desiccation but allows free gaseous exchange. If the object is to obtain single cell clones, the plates are viewed with a dissecting microscope and the positions of the individual single cells marked by circling with a fine marker pen on the outside of the dish. The cultures are then incubated at 25° C in the dark. The marked cells are observed at intervals to see whether they have divided to give rise to colonies, and these are transferred to fresh medium when they reach a suitable size.

If quantitative information regarding the ability of single cells and small cell aggregates to grow is required, the method is modified as follows. The cell density in the suspension is carefully adjusted to the required level, mixed with agar medium and dispensed as described above. A grid is drawn on the under surface of the petri dish to facilitate counting and the numbers of single cells and small cell aggregates (cell units) per plate estimated before incubation. After 21 days at 25° C the plates are re-examined and the number of colonies formed per plate estimated. From this the plating efficiency is calculated.

$$\text{Plating efficiency} = \frac{\text{Number of colonies per plate}}{\text{Number of cell units per plate}} \times 100$$

Many of the colonies raised by this method come from small cell aggregates rather than single cells. However, it is reasonable to assume that such aggregates themselves originally came from single cells.

5.4   Growth of cells in micro-chambers

Hildebrandt and co-workers in 1960 attempted to culture single cells of a tumorous hybrid tobacco plant (*Nicotiana tabacum* × *N. glutinosa*). Using a micro-chamber (Fig. 5–3) sealed with inert mineral oil they found that a proportion of the single cells divided. They also observed that those cells which divided and then became senescent induced cell division in other, previously quiescent, cells. Moreover, single cells divided when placed in a micro-chamber containing conditioned media, i.e. media in

44

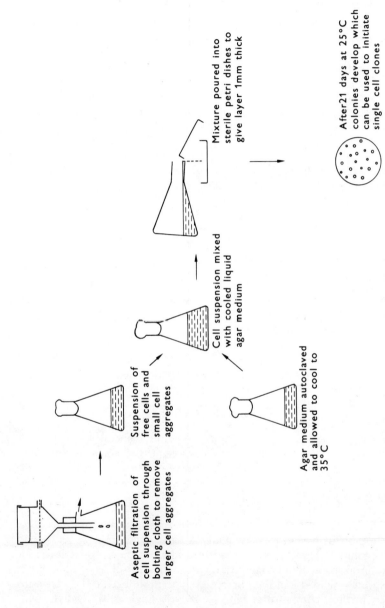

Aseptic filtration of cell suspension through bolting cloth to remove larger cell aggregates

Suspension of free cells and small cell aggregates

Cell suspension mixed with cooled liquid agar medium

Agar medium autoclaved and allowed to cool to 35°C

Mixture poured into sterile petri dishes to give layer 1mm thick

After 21 days at 25°C colonies develop which can be used to initiate single cell clones

Fig. 5-2 Procedure for obtaining single cell clones using a petri dish plating technique.

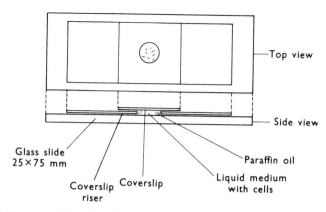

Fig. 5–3 Microchamber used to observe the growth of single cells.

which callus and suspensions had previously been cultured. Colonies derived from such cells could then be subcultured on to fresh medium to give rise to single cell clones.

5.5 Factors affecting growth

The work outlined above shows clearly that single cells have more exacting nutrient requirements than comparable callus and suspension cultures. For example, the induction of division of single cells using the paper raft technique indicates that essential nutrients from the callus mass as well as from the medium diffuse through the paper barrier. The requirement for nutrients from living tissues can also be demonstrated when callus masses are plated on to agar medium seeded with single cells (Fig. 5–4); cells first begin to divide in regions near the callus mass.

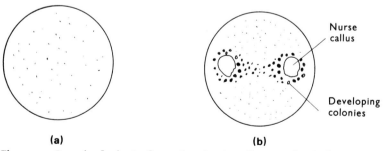

Fig. 5–4 Growth of colonies from a low density cell suspension in the presence of callus tissues. (a) Petri dish inoculated with low density suspension of *Acer* cells—no colonies develop. (b) Petri dish inoculated with low density suspension plus nurse callus—colonies grow near to nurse calluses only.

Further evidence that essential metabolites are liberated from cells comes from studies using the agar plating technique, where it has been shown that there is a critical density below which plated single cells and small aggregates will not divide and develop into colonies.

In order to explain these findings it has been postulated that the plasmalemmae of cultured cells are leaky and allow diffusion of metabolites into the surrounding medium. Each cell is presumed to be capable of synthesizing all the metabolites required, but critical endogenous levels must be established before growth can proceed. These critical endogenous levels are reached in high density cell populations, but not in low density populations where the metabolites diffuse out into a relatively large volume of culture medium. The actual metabolites involved have still to be properly identified, but there is some evidence to implicate cytokinins, gibberellins, ethylene, amino acids and carbon dioxide. For example, petri plates inoculated with low density suspensions of *Convolvulus* cells require cytokinins and certain amino acids before growth occurs, while these supplements are not required for the growth of comparable callus tissues. Similarly, sycamore cells at low plating densities require cytokinin, gibberellin and certain amino acids for growth, while high density suspensions and callus cultures do not.

These studies, in addition to confirming that isolated single cells have more exacting nutritional requirements than comparable callus and suspension cultures, also suggest that there are mutually beneficial interactions between individuals in a growing cell population.

5.6   Characteristics of single cell clones

The paper raft nurse technique, the petri dish plating technique and micro-chambers have all been used successfully to obtain single cell clones. Clones which have been derived from established callus and suspension cultures have shown the heterogeneity of such cultures. Those derived from a particular callus or suspension culture may differ in growth rate, nutritional requirements, texture, pigmentation and ability to undergo organogenesis and embryogenesis (to form roots, buds and embryoids). Furthermore, when secondary single cell clones are obtained from a primary single cell clone considerable differences between the isolates are again encountered. It follows that the heterogeneity must have arisen spontaneously in the primary single cell clone. It is likely that some of these variations between clones are a consequence of the chromosomal irregularities which occur (see Chapter 3, section 3.4), but other factors may also be involved.

6 The Isolation and Culture of Plant Protoplasts

6.1 Introduction

The techniques of plant tissue culture have recently been extended to include the culture of naked plant protoplasts (Fig. 6–1), obtained by

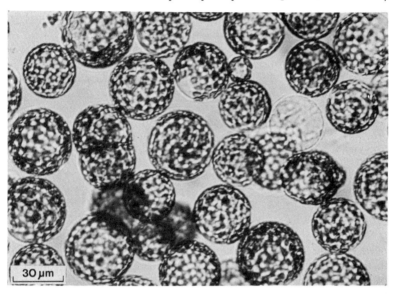

Fig. 6–1   Naked protoplasts isolated from the mesophyll cells of a tobacco leaf. (Photograph kindly supplied by Dr. J. B. Power.)

removing the cell wall. Isolated protoplasts are potentially valuable experimental material for the isolation of undamaged organelles and macromolecules and for the investigation of physiological phenomena such as the transporting properties of the plasmalemma and cell wall regeneration. Although important, these uses of protoplasts are secondary compared with their exciting potentials in the fields of somatic cell hybridization and plant breeding, and in the studies of virus infection.

6.2 Isolation

The earliest attempts to isolate protoplasts can be traced back to Klerker in 1892 and Kuster in 1940. Their methods consisted of plasmolysing cells and cutting the tissue to release the protoplasts during de-plasmolysis. This method, although effective, provides insufficient protoplasts for most experimental purposes. In 1960 Cocking introduced a much more efficient method involving the use of cell wall degrading enzymes.

The most commonly used enzymes have been purified preparations of cellulases, hemicellulases and pectinases from fungi. A number of procedures have now been worked out for isolating protoplasts from various tissues (leaves, pollen, callus, suspension cells) of a large number of species. One of these is illustrated in Fig. 6–2. For most purposes it is essential to prepare the protoplasts under aseptic conditions, since the media used for isolation favour microbial growth. The sterilized tissues or cells are first placed in a medium containing 13% mannitol, where they become slightly plasmolysed. They are then transferred to an enzyme mixture containing cellulases in combination with hemicellulases and/or pectinases and incubated for 4–6 hours at 22–25° C. Finally the protoplasts are washed free of enzymes and suspended in a small volume (10 ml) of medium containing 10–13% mannitol. The newly released protoplasts are fragile and must be stabilized in media where the osmotic properties are closely controlled. Sorbitol and mannitol are the most commonly used osmotic stabilizers.

6.3 Culture

Many of the uses of protoplasts depend on their ability to regenerate new cell walls and recover the capacity for division and growth. Two methods have been employed, one of which is to place droplets of protoplast suspension in small petri dishes which are then sealed with plastic film, while the other is to embed the protoplasts in a thin layer of agar in petri dishes. The latter method closely resembles the petri dish plating techniques used for isolating single cell clones (Chapter 5, section 5.3). Usually the basic nutrient requirements for cell wall regeneration and growth are similar to those of other plant cells, but the level of some components such as calcium and carbon source are often more critical. The osmotic properties of the cultures are adjusted by the addition of sorbitol or mannitol in a range varying from 0.3–0.8 M depending on the source of the protoplasts.

When the protoplasts are washed free of the wall degrading enzymes and placed in a favourable medium cell wall regeneration takes place almost immediately. The newly formed walls can be shown by plasmolysis or by staining with a fluorescent reagent such as 'Calcofluor White', but the detailed structure is best demonstrated by electron microscopy. If the medium contains the appropriate growth factors the regenerated cells will divide to form small cell aggregates or colonies. Sustained cell

49

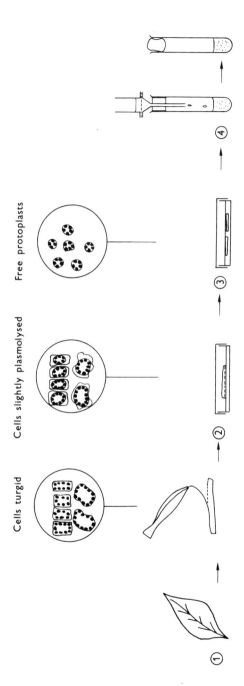

Fig. 6–2 Procedure for isolating protoplasts from the mesophyll cells of a tobacco leaf. Stage 1: leaf surface is sterilized with 70% ethanol and 2% sodium hypochlorite. The epidermis is stripped from the upper surface to expose mesophyll cells. Stage 2: the leaf is then placed, stripped surface downwards, in a medium containing 13% mannitol for 3 hours to plasmolyse mesophyll cells. Stage 3: the leaves are cut into sections and transferred to an enzyme mixture containing cellulase and pectinase for 4 hours at 22° C. Stage 4: the liberated protoplasts are filtered through a coarse wire gauze to remove most of the debris. Protoplasts are then collected by centrifugation, washed and finally suspended in a small volume of culture medium containing 13% mannitol.

division and growth have now been observed in regenerated 'protoplast' cells of a number of species including tobacco, carrot, soybean and petunia.

The cells regenerated from protoplasts possess the same properties for organogenesis as the cell cultures derived directly from somatic tissues. Normal plants have now been raised from protoplasts derived from leaf mesophyll cells of tobacco and cultured cells of carrot.

6.4 Protoplasts and viral infections

Investigations of the penetration and multiplication of viruses in plant cells have been hampered by the lack of suitable experimental systems. This is because the only method of inoculation available to the virologist has been to rub the surface of the host leaf with a viral suspension containing an abrasive. Such treatment damages the cell walls, allowing the entry of the virus particles through the exposed membranes. This technique is inefficient, cannot be quantified and gives inconsistent results. It is clear that much of the problem stems from the presence of the cell wall, and it would be expected that naked protoplasts would provide a much better system for studying viral infection. Indeed it has been shown that 30–70% of the protoplasts from tobacco leaf cells become infected when incubated with tobacco mosaic virus (TMV). The infectivity is greatly enhanced by the presence of poly-L-ornithine, which is known to promote the uptake of proteins in mammalian cells. It is reasonable to assume that the compound has a similar function in plant protoplasts and assists viral entry. After entering the protoplast the viruses multiply rapidly, the rate of multiplication being highest 8–22 hours after infection. The significant features of the system are the high percentage of infection and the efficiency of multiplication; it has been estimated that the efficiency is a hundredfold greater than can be achieved by the conventional mechanical inoculation of tobacco leaves. The protoplast system is now being extended to studies of infection by a number of other viruses and may soon be extended further to include studies of nitrogen fixing organisms and plant pathogenic bacteria and fungi.

6.5 Protoplast fusion and somatic hybridization

Certain intra- and inter-specific hybrid plants are impossible to produce because of incompatibility factors which prevent crosses by normal pollination procedures. The most intriguing aim of those working with isolated protoplasts is to produce such plants by somatic cell hybridization. Clearly if this objective is realized it should be possible to transfer 'useful' genes (for disease resistance, nitrogen fixation, rapid growth rate, protein quality and frost hardiness) from one species to another and thereby widen the genetic base for plant breeding. It does not require much imagination to appreciate the tremendous effect this could have on agricultural economy. An outline of the procedures which must be developed before successful somatic hybridization can be achieved is given below.

(a) Isolation of protoplasts from the appropriate species.
(b) Fusión of protoplasts from the different species to produce viable heterokaryons (cells containing nuclei from different sources).
(c) Wall regeneration by heterokaryotic cells.
(d) Fusion of nuclei within the heterokaryons to produce hybrid cells.
(e) Division of hybrid cells to form colonies.
(f) Selection of desired hybrid colonies.
(g) Induction of organogenesis in the hybrid colonies.
(h) Raising mature plants from regenerated shoots or embryoids.

Although all these steps are theoretically possible, there are considerable practical difficulties. The first step, that of obtaining the protoplasts of the appropriate species, has been largely overcome and there are now methods available for isolating protoplasts from a wide range of plants. Progress is also being made towards the development of procedures to facilitate adhesion and subsequent fusion of protoplasts from different species. It has been shown that centrifugation, osmotic shocks and the addition of sodium nitrate or polyethylene glycol encourage protoplast fusions. Already fusions have been accomplished between protoplasts of widely different species such as soybean and wheat and rice and wheat. Following fusion these heterokaryons regenerate walls in a way similar to other protoplasts. In a few cases cell wall regeneration has been followed by the synchronous mitosis of the two nuclei within the heterokaryon. So far, however, nuclear fusion has not been observed. If we assume that nuclear fusion will eventually be achieved and that the resulting hybrid cells will divide to form colonies it will still be necessary to find effective ways of identifying and selecting the desired hybrid colonies from the others present. This becomes a significant problem when it is appreciated that the desired fusions may occur at very low frequencies. At present a few morphological characters such as pigmentation, and cytological characters such as chromosome structure, are being used to identify cell fusions, but it is clear that these will be inadequate for the selection of many hybrid cells. There is a serious need to develop biochemical mutant markers if the present difficulties of selection are to be overcome. Since hybrid cells from widely different species have not been reported it is not possible to say whether or not colonies of hybrid cells can be induced to undergo organogenesis, but it seems probable that difficulties may be encountered when the parent species are unrelated. Thus there is a long

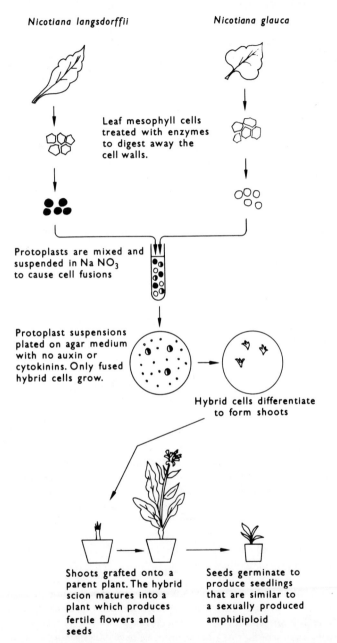

Nicotiana langsdorffii   *Nicotiana glauca*

Leaf mesophyll cells treated with enzymes to digest away the cell walls.

Protoplasts are mixed and suspended in Na NO₃ to cause cell fusions

Protoplast suspensions plated on agar medium with no auxin or cytokinins. Only fused hybrid cells grow.

Hybrid cells differentiate to form shoots

Shoots grafted onto a parent plant. The hybrid scion matures into a plant which produces fertile flowers and seeds

Seeds germinate to produce seedlings that are similar to a sexually produced amphidiploid

Fig. 6–3 Procedure for obtaining somatic hybrid plants from *Nicotiana glauca* and *N. langsdorffii*. (Adapted from SMITH, H. H., *BioScience* 24, 269, 1974.)

way to go before somatic hybridization by protoplast fusion techniques becomes a standard method for use by plant breeders.

One method to test the feasibility of somatic cell hybridization has been to use closely related species within the genus *Nicotiana* (tobacco) as experimental models (Fig. 6–3). Carlson has prepared protoplasts from *N. glauca* and *N. langsdorffii* species which, when crossed sexually, produce hybrids which readily form spontaneous genetic tumours. It is well known that callus tissues from the tissues of this hybrid have no requirements for auxin or cytokinin. This information indicates a very effective way of selecting hybrid cells which may result from protoplast fusions. Carlson treated a mixture of the protoplasts from the two species with sodium nitrate and up to 25% of the protoplasts were observed to be involved in fusions. The cells were allowed to regenerate walls and form callus on a selection medium lacking auxin and cytokinin. Eventually shoots were formed by the callus and were grown to maturity by grafting onto cut stems of *N. glauca*. Three of the plants regenerated in this way were found to be amphidiploid hybrids comparable in every way to the sexual hybrids. Thus, although detailed observations of nuclear fusion were not made, the results suggest that somatic hybridization between these two closely related species had been achieved. However, it should be appreciated that the selection procedure was unique and could not have been anticipated without a prior knowledge of the properties of the hybrid obtained by normal sexual procedures. It is interesting to note that Melchers and his colleagues have recently reported the production of hybrids from the fusion of haploid protoplasts of two different chlorophyll deficient varieties of *N. tabacum*. A regenerated plant with the diploid number of chromosomes was isolated which was indistinguishable from comparable sexual hybrids.

It may be concluded that there is good evidence that protoplast fusion techniques can be used to obtain somatic hybrids between closely related species, but that there are many obstacles to be overcome before it will be achieved between unrelated plants.

Another possible way of achieving gene transfer between unrelated species would be to feed DNA isolated from one species to the protoplasts of another. So far there is little evidence that this method will be a realistic possibility in the near future.

7 Résumé

In the foregoing six chapters we have described and characterized the many different kinds of cell and organ cultures, and have indicated how they are being used to solve a variety of research problems. As anticipated by Haberlandt, tissue cultures have made important contributions to studies of plant morphogenesis. Indeed the study of morphogenetic problems has provided the foundations for the development of all the techniques in current use. The highly successful techniques of single cell cloning, cell suspension culture and induction of organogenesis and embryogenesis in callus have all developed from discoveries made during investigations of morphogenesis.

Although studies of tissue cultures are still furthering our understanding of plant development, it has become apparent in recent years that they have enormous potential in other fields of study. For example the large scale suspension cultures grown under strictly controlled conditions in the absence of micro-organisms provide excellent systems for many physiological and biochemical projects as well as providing a system for producing commercially important plant metabolites such as alkaloids. Similarly, the prospect of having haploid cell suspensions which can be 'single cell cloned' offers an attractive system for studying the biochemical genetics of plant cells.

Furthermore, the special properties of some cultures are of particular significance in plant breeding. The production of numerous haploid plantlets from anther cultures and pollen offers a quick and convenient way of obtaining the homozygous plants necessary for breeding programmes. Breeding may be further speeded up by using tissue culture methods (callus and explants from apices) for vegetatively propagating valuable stock lines. Finally, there is the exciting possibility of using protoplast fusion techniques for the somatic hybridization of unrelated species.

In plant pathology tissue culture techniques have been extremely valuable in the study of tumour diseases, and are currently being assessed as suitable systems for studying host-parasite relationships. Already in this field protoplasts are proving to be very useful in studies of virus penetration and multiplication, while other tissue culture methods are being used to free vegetatively propagated plants of debilitating viruses.

The formation of plantlets in callus cultures and explants is also being exploited as a means of propagating commercially valuable plants such as orchids.

8 How to Grow Cultures

It is not difficult to grow and experiment with plant tissue cultures. In this chapter we describe the apparatus and chemicals that are required, and the basic methods that are employed. All operations can be carried out in a simply equipped laboratory.

8.1 Sterility

The maintenance of sterility is of fundamental importance because all media used contain nutrients which support the growth of many bacteria and fungi. These would soon outgrow and destroy the slower growing plant tissues if precautions were not taken to exclude them. Even slow growing contaminants are undesirable because they may produce toxins or even stimulants which affect the growth of plant cells. For these reasons plant organs must be surface sterilized chemically at the time of culture initiation, whilst instruments, culture vessels and media must be sterilized before use. All culture manipulations must be carried out aseptically in a specially designed inoculating hood.

8.2 Equipment

8.2.1 Culture vessels and glassware

All culture vessels and glassware used in the preparation of media should be of borosilicate glass (e.g. 'Pyrex' or 'Monax'). Soda glass is usually toxic to tissue cultures, particularly isolated roots. Glassware should be cleaned by soaking overnight in strong detergent, brushing and washing well in tap water and then rinsing with at least two changes of distilled or deionized water. It should be dried in an oven and then stored in a dust-proof cupboard or drawer.

Many different kinds of vessels may be used for growing cultures. We recommend 100 ml, wide mouth conical flasks (Erlenmeyer flasks) for both root and callus cultures, but if space is a problem, callus cultures can be grown quite successfully in large test tubes (25 × 150 mm). Culture vessels must be fitted with closures which exclude microbial contaminants yet allow free gas exchange. These may be tightly rolled plugs of non-absorbent cotton wool wrapped in muslin. When in position the exposed part of each plug and the rim of the culture vessel should be covered by a cap of aluminium foil. This will keep the plug and the vessel rim free of dust, and will protect the plugs from wetting during autoclaving. Any plugs which become wetted or contaminated with media should be discarded.

In addition to the culture vessels the following glassware will be needed for making media and preparing tissues for culture: graduated pipettes (1.0 ml and 10.0 ml); measuring cylinders (100 ml and 1 l); conical flasks (100 ml, 1 l and 2 l); beakers (250 ml); a filter funnel; and petri dishes (9 cm diameter).

8.2.2 Autoclave

An autoclave is used for melting agar and for sterilizing distilled water, petri dishes and culture vessels containing media. It operates on the same principle as a pressure cooker, the items to be sterilized being heated by steam at a pressure of 15 lb/in^2 (103 kPa) (121° C) for 10 minutes. Small portable electric or gas autoclaves are sold by A. Gallenkamp and Co. Ltd. (Fig. 8–1). However, even simple models like these are very

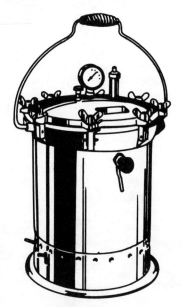

Fig. 8–1 A simple portable electric autoclave.

expensive, and if your laboratory does not possess one you should use a large domestic pressure cooker instead.

8.2.3 Instruments

These should be of stainless steel or nickel plated. You will need: a sharp scalpel (surgical scalpels with disposable blades are ideal); scissors with fine, curved points (11.5 cm); forceps with fine points (15 cm straight ended and 12.5 cm with curved points); a bacteriological-type loop of stainless steel or nichrome wire (gauge 24) fitted into a long-handled holder. A spirit burner or gas micro-burner will be needed for the flame sterilization of instruments.

8.2.4   Inoculating hood

In a research laboratory all aseptic manipulations and transfers are carried out in an inoculating room or at a hooded bench fed with moving sterile air. However, a simple inoculating hood such as that supplied by Harris Biological Supplies Ltd, will be found adequate for most purposes. If you are unable to purchase a hood such as this you may be able to make one from sheet metal and glass, using the design shown in Fig. 8–2. In the absence of a hood, aseptic procedures should be carried out in a small laboratory with windows and doors closed.

The internal surfaces of the transfer hood should be kept very clean and wiped frequently with a cottonwool pad soaked with methylated spirit.

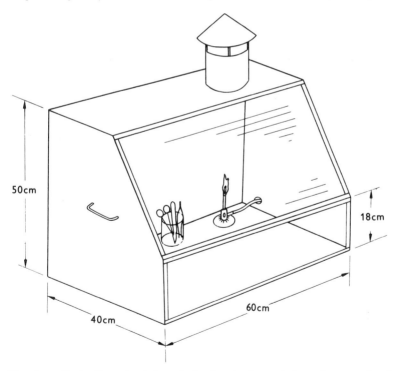

Fig. 8–2   Design for a simple inoculating hood to be made from sheet metal and glass.

Instruments (scalpels, scissors, etc.) should be kept standing in a beaker of methylated spirit in a corner of the hood and should be sterilized before each tissue-handling operation by flaming over a conveniently placed gas or spirit burner and then being allowed to cool. [*Caution—methylated spirit is highly inflammable; do not place hot instruments in spirit.*] The burner is also used to flame the mouths of sterile culture vessels each time that plugs are removed for inoculation or tissue transfer. Simply pass the mouth of the vessel through the flame once, taking care not to heat the glass excessively. It is well to remember that speed is an important factor in aseptic manipulations. Think of what you will need before you start work, and then position instruments, culture vessels, etc. in the hood so as to ensure that plant materials and media are exposed for the minimum time.

8.2.5   *Incubator*

Most plant tissue cultures grow best at constant temperatures with low intensity light. A small room lit by a single fluorescent tube and maintained at 25° C is ideal for incubation. However, cultures will grow very well in a dark cupboard or incubator at temperatures anywhere between 20° C and 28° C. At lower temperatures growth will be slow, while at higher temperatures the tissues may be damaged. Cultures should not be subjected to rapid changes of temperature.

8.3   Initiation of isolated root cultures of tomato

We recommend the use of seed of the tomato variety Sutton's Potentate—Best of All.

8.3.1   *The culture medium*

The following medium, which will support good growth of isolated roots of tomato, is a modification of a formula originally devised by White.

8.3.2   *Preparation of the culture medium*

Tissue cultures are very sensitive to toxic chemical contaminants in the culture medium. For this reason all solutions should be prepared with glass distilled water and all chemicals should be of the highest grade of purity available. Inorganic chemicals should, wherever possible (always in the case of $FeCl_3.6H_2O$), be of 'Analar' or similar grades. However, if these are not available, general purpose reagents will probably be adequate.

It would be very time consuming to weigh out each constituent every time a batch of medium were required. It is convenient, therefore, to store most of the compounds as concentrated stock solutions as follows.

Solution A—Inorganic salts (excluding the iron source). Make up at 10

Constituents	Content/litre of medium	
Inorganic salts:		
Calcium nitrate, $Ca(NO_3)_2.4H_2O$	290	mg
Magnesium sulphate, $MgSO_4.7H_2O$	730	mg
Potassium chloride, KCl	65	mg
Potassium nitrate, KNO_3	80	mg
Sodium sulphate, $Na_2SO_4.10H_2O$	450	mg
Sodium dihydrogen phosphate, $NaH_2PO_4.2H_2O$	22	mg
Boric acid, H_3BO_3	1.5	mg
Copper sulphate, $CuSO_4.5H_2O$	0.02	mg
Manganous chloride, $MnCl_2.4H_2O$	6.0	mg
Molybdic acid, H_2MoO_4	0.0017	mg
Potassium iodide, KI	0.75	mg
Zinc sulphate, $ZnSO_4$	2.6	mg
Iron source:		
Ferric chloride, $FeCl_3.6H_2O$	3.1	mg
Sodium ethylenediaminetetra-acetate, EDTA	8.0	mg
Vitamins and other additives:		
Aneurine hydrochloride (thiamine HCl)	0.1	mg
Pyridoxine hydrochloride	0.1	mg
Nicotinic acid	0.5	mg
Glycine	3.0	mg
Carbon source:		
Sucrose	20	g
pH 4.8		

times the final strength of the medium and store in a refrigerator. To make 1 litre of this stock solution dissolve the salts, *one at a time*, in 750 ml of distilled water and then make up to volume.

Solution B—Iron source (i.e. $FeCl_3.6H_2O$ and EDTA). Make up at 10 times the final strength of the medium and store in a refrigerator (approximately 5°C).

Solution C—Vitamins and glycine. Make up at 1000 times the final strength of the medium, divide into aliquots of 5 ml and store in a deep freeze. If a deep freeze is not available, vitamin solutions must be freshly made up each time they are required.

It is not sufficient to mix the stock solutions at random when making up a batch of medium. To avoid precipitation of the salts the following procedure should be strictly adhered to.

To make 1 litre of medium:

(1) Dissolve 20 g of sucrose in 600 ml of distilled water contained in a 2 litre flask.

(2) Add 100 ml each of stock solutions A and B and 1 ml of stock solution C (thawed) in the order stated. Take care to mix the medium well before each addition.

(3) Pour into a 1 litre measuring cylinder, make up to 1 litre with glass distilled water, return to the 2 litre flask and again mix well.

(4) Adjust the pH to 4.8–5.0 with the aid of a few drops of 0.1 N Na OH or 0.1 N HCl. The pH of the medium is very important and this operation should be carried out with great care. If your laboratory does not possess a pH meter you should use narrow range pH test papers (range, pH 4–6).

(5) Dispense the medium into culture flasks (50 ml medium/flask), insert cotton wool plugs, cover with foil caps and sterilize by autoclaving at 15 lb/in² (103 kPa) for 10 minutes.

8.3.3 Initiation of tip cultures

Isolated root cultures of tomato are initiated from the root tips of aseptic seedlings. These are produced by surface sterilizing the seeds in a chlorine solution to destroy the spores of bacteria and fungi (internal tissues are usually free of micro-organisms), and then germinating them in sterile petri dishes.

To initiate 20 cultures the following items will be required:

(a) About 60 seeds of the tomato variety Sutton's Potentate—Best of All.

(b) An empty, sterile, 100 ml conical flask fitted with a cotton wool plug.

(c) About 50 ml of 80% ethanol.

(d) A solution of sodium hypochlorite in water (1.6% available chlorine).

(e) About 500 ml of sterile distilled water (i.e. autoclaved at 15 lb/in² (103 kPa) for 10 minutes in a flask fitted with a cotton wool plug and allowed to cool).

(f) Six sterile 9 cm petri dishes, each containing 2 sterile filter papers wetted by adding 10 ml of distilled water. Wrap the dishes containing the filter papers in aluminium foil and autoclave at 15 lb/in² (103 kPa) for 10 minutes.

(g) 20 culture vessels (100 ml wide mouth conical flasks), each containing 50 ml of sterile culture medium.

(h) One 500 ml beaker.

(i) A sharp scalpel, forceps, scissors and a wire loop.

The procedure for initiating the cultures is set out in Fig. 8–3. All manipulations should be carried out in an inoculating hood.

8.3.4 Initiation of clones (see section 2.2)

The number of isolated root cultures may be increased, or the existing cultures maintained, by transfer of root tips or segments to fresh medium

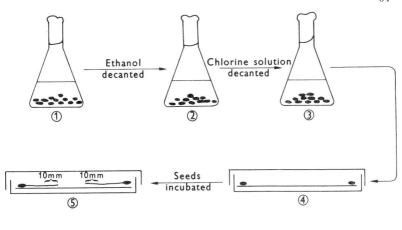

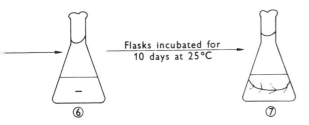

Fig. 8–3 Procedure for the initiation of isolated root cultures of tomato.

(1) Place the seeds in the plugged 100 ml flask and cover with 80% ethanol. Leave for 1 minute.

(2) Decant the ethanol into the 500 ml beaker and replace with the chlorine solution. Replace the plug and leave for 10 minutes. Ensure that the seeds sink in the solution by agitating the flask.

(3) Decant the chlorine solution into the beaker and wash the seeds with 3 changes of sterile distilled water (still in flask).

(4) Using flamed forceps transfer 6–10 seeds to each sterile petri dish containing wet filter paper.

(5) Incubate the dishes in the dark until the radicles are 30–40 mm long (about 5 days at 25° C).

(6) Excise 10 mm apical tips of the roots using a sharp flamed scalpel and transfer them singly with a flamed wire loop to the flasks of culture medium. Great care should be taken to avoid damaging the root tips during this operation.

(7) Incubate the cultures for 10 days at 25° C. After this time each apical tip should have developed into a root, 100–200 mm long, with numerous laterals. If roots do not grow and the medium is cloudy it is likely that the cultures are contaminated with fungi or bacteria.

after 10 days, and then every 7 days. The procedure is as follows (see Fig. 8–4).

Transfer a 10-day-old tip culture aseptically, using a flamed wire loop, to a sterile petri dish containing sterile medium. Next, using flamed scissors, cut the main axis of the root as shown in Fig. 8–4 to provide a number of 'sector initials', each consisting of a portion of the main axis of the root bearing 4 or 5 laterals. Transfer these sector initials individually to flasks of fresh culture medium and incubate at 25° C for 7 days. The 'sector cultures' which result are composed of the original short lengths of main axis bearing numerous lateral roots which have themselves developed further laterals. Such sector cultures can be transferred to petri dishes of medium and cut up, as shown in Fig. 8–4, to yield 4–5 root tips (10 mm long) which can be used to initiate further tip cultures for experimental use, and a number of sector initials which can be grown on and used to provide tips for later experiments.

8.3.5   Suggested experiments with isolated roots

Root tips obtained either from sterile seedlings or from clonal cultures may be used.

(1) Investigate the vitamin requirements of isolated tomato roots by comparing growth from 10 mm apical tips in the complete medium and in media from which one or more of the vitamins have been omitted. Growth may be assessed by measuring the length of the main axes of the roots and by counting the number of laterals and measuring their total length after 7 days of incubation at 25° C. Root growth may also be assessed on a fresh or dry weight basis.

(2) Compare the ability of various carbohydrate sources to support the growth of tomato roots. We suggest that you compare sucrose (2%), glucose (1%), fructose (1%), glucose (1%) + fructose (1%). Growth may be assessed as described above.

(3) Determine the effect of auxin on the growth of tomato roots. We suggest that you use the synthetic auxin 1-naphthaleneacetic acid (NAA) at the following concentrations: 10^{-5}, 10^{-4}, 10^{-3} and 10^{-2} mg/l of medium. A stock solution of NAA can be made as follows. Dissolve 10 mg of NAA in 2 ml of 0.1 N NaOH, dilute to 1 litre with glass distilled water and adjust the pH to 5.5 with 0.1 N HCl. Take 10 ml of this solution, discarding the rest, and again make up to 1 litre with glass distilled water. If 10 ml of this stock is used in making 1 litre of medium it will give a final concentration of 10^{-2} mg NAA/litre; 1 ml will give 10^{-3} mg/l and so on.

(4) Investigate the ability of the modified White's medium (8.3.1) to support the growth of isolated roots of other plant species. We suggest that you try carrot, pea and white mustard.

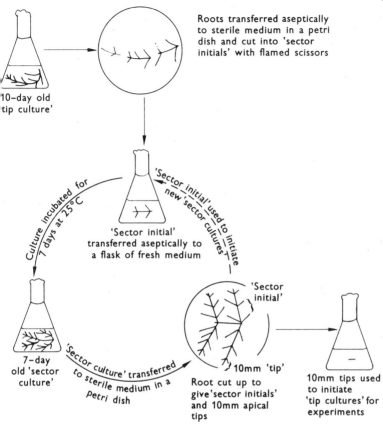

Fig. 8-4 Procedure for the maintenance of isolated tomato roots in continuous culture.

8.4 Initiation of callus cultures from the tap root of carrot

We suggest that you use mature tap roots of the variety Chantenay Red Cored, although other varieties will give satisfactory results.

8.4.1 The culture medium

The requirements for growth of callus tissue are different from those for the growth of isolated root cultures. The following medium is based on that described in 8.3.1, but contains hormone substitutes and extra salts and is solidified with agar. It is suitable for growing callus cultures of carrot and many other species.

Constituents	Content/litre medium	
Inorganic Salts:		
Ammonium sulphate, $(NH_4)_2SO_4$	790	mg
Calcium nitrate, $Ca(NO_3)_2.4H_2O$	290	mg
Magnesium sulphate, $MgSO_4.7H_2O$	730	mg
Potassium chloride, KCl	910	mg
Potassium nitrate, KNO_3	80	mg
Sodium nitrate, $NaNO_3$	1800	mg
Sodium sulphate, $Na_2SO_4.10H_2O$	450	mg
Sodium dihydrogen phosphate, $NaH_2PO_4.2H_2O$	320	mg
Boric acid, H_3BO_3	1.5	mg
Copper sulphate, $CuSO_4.5H_2O$	0.02	mg
Manganous chloride, $MnCl_2.4H_2O$	6.0	mg
Potassium iodide, KI	0.75	mg
Zinc sulphate, $ZnSO_4.7H_2O$	2.6	mg
Molybdic acid, H_2MoO_4	0.0017	mg
Iron source:		
Ferric chloride, $FeCl_3.6H_2O$	3.1	mg
Sodium ethylenediaminetetra-acetate (EDTA)	8.0	mg
Vitamins, etc:		
meso-inositol	100	mg
Glycine	3.0	mg
Aneurine hydrochloride (thiamine HCl)	0.1	mg
Pyridoxine hydrochloride	0.1	mg
Nicotinic acid	0.5	mg
Hormone substitutes:		
2, 4-dichlorophenoxyacetic acid (2,4-D)	0.15	mg
6-furfurylaminopurine (kinetin)	0.15	mg
Carbon source:		
Sucrose	20	g
Agar:		
Oxoid No. 3 (or similar)	7	g

The volume is made up to 1 litre with glass distilled water and the pH adjusted to approximately 5.5.

8.4.2 Preparation of the medium

The procedures for making stock solutions and for preparing 1 litre of medium are the same as described in 8.3.2, with the following additions.

(a) Separate stock solutions must be made up for the plant hormone substitutes and aliquots added to the medium after the vitamins. These stocks may be made in the following way.

2,4-D stock solution. Dissolve 30 mg of 2,4-D in about 2 ml of 0.1 N NaOH, dilute to 100 ml with distilled water, and adjust the pH to 5.5 with 0.1 N HCl. Use 0.5 ml of stock per litre of medium.

Kinetin stock solution. Dissolve 7.5 mg of kinetin in a minimum quantity (not more than about 2 ml) of 0.1 N HCl, dilute to 1 l with distilled water and adjust the pH to 5.5 with a few drops of 0.1 N NaOH. Use 20 ml of stock per litre of medium.

(b) The agar powder should be added to the medium after adjusting the pH to 5.5.

(c) The medium should be autoclaved in bulk in a 2 litre flask at 15 lb/in^2 (103 kPa) for 1 minute to dissolve the agar. It should then be mixed thoroughly and dispensed, whilst still hot, into culture flasks (50 ml/flask) or culture tubes (20 ml/tube). Finally, the culture vessels should be fitted with cotton wool plugs and foil caps and sterilized by autoclaving at 15 lb/in^2 for 10 minutes.

8.4.3   The initiation of callus cultures

The internal tissues of the carrot root are usually free of contaminants, although many micro-organisms are present on the surface. Before callus cultures can be initiated it is therefore necessary to surface sterilize the roots with mercuric chloride.

To initiate 20 cultures you will need:

(a) A large carrot root, 10–14 cm long. Do not use washed carrots from the supermarket as these frequently contain bacterial contaminants introduced into the tissues through minor injuries inflicted during the washing process.

(b) A sterile, 250 ml beaker fitted with a petri dish lid.

(c) A 0.1% solution of mercuric chloride. [*Note—this substance is toxic and should be handled with care.*]

(d) About 750 ml of sterile distilled water contained in a flask plugged with cotton wool.

(e) Sterile petri dishes.

(f) 20 culture flasks, each containing 50 ml of the sterilized callus culture medium.

(g) A sharp scalpel and forceps.

(h) A large beaker for waste solutions.

The procedure for initiating the cultures is set out in Fig. 8–5. All manipulations should be carried out in an inoculating hood or room.

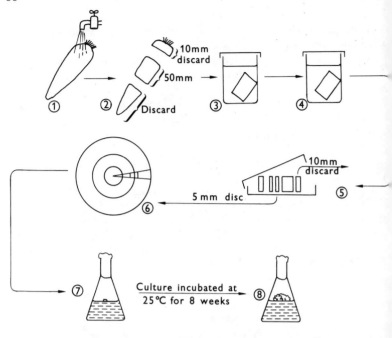

Fig. 8–5 Procedure for the initiation of callus cultures of carrot.

(1) Wash the carrot root in running tap water, taking care not to damage the skin.

(2) Cut a 50 mm long segment of tissue from the root as shown.

(3) Place the 50 mm segment of tissue into the sterile beaker, cover with mercuric chloride solution and leave for 30 minutes for surface sterilization to occur.

(4) Decant the mercuric chloride solution into the waste beaker, replace with sterile distilled water and rinse the tissue piece well. Repeat the rinsing operation with three changes of sterile water to remove all traces of mercuric chloride.

(5) Using flamed forceps transfer the tissue segment to a sterile petri dish and cut off a 10 mm slice from each end with a flamed scalpel. These should be discarded. Again using a flamed scalpel slice the remainder of the segment into discs 5 mm thick, and transfer these singly to clean sterile petri dishes.

(6) Cut cubes of tissue, about 5 mm^3, from the region of the cambium.

(7) Place the cubes of tissue singly on the surface of the culture medium in the culture flasks and incubate at 25° C.

(8) After 14–21 days the callus tissue will have started to develop, and by 6 weeks the cultures should be well advanced.

After 6–8 weeks it will be necessary to transfer the callus tissue to fresh medium. Remove each clump of tissue to a sterile petri dish using flamed forceps. Cut into small pieces (about 100 mg) with a scalpel and transfer these singly to flasks of fresh medium.

8.4.4   Suggested exercises with callus cultures

(1) Make a careful microscopic investigation of the initiation of carrot callus from the normal root tissue. Describe the range of cell types to be found in callus tissue.

(2) Study the differentiation of roots, shoots and plantlets on carrot callus by omitting 2,4-D and kinetin from the culture medium, both separately and together. Can you devise a way of growing the 'callus plantlets' to full maturity?

(3) Attempt to grow callus cultures from the tissues of other Angiosperms. We suggest that you try potato tubers, the swollen hypocotyls of turnip and beetroot and the stems of sunflower, tomato and willow. The potato, turnip and beetroot can be handled in a similar way to the carrot. The stem tissues of sunflower, tomato and willow, however, require different techniques. Choose young stems, about 5 mm thick, and surface sterilize 30 mm internode segments in a chlorine solution (see 8.3.2 for preparation) for 10 minutes. Wash twice in sterile distilled water, cut discs of tissue, about 5 mm thick, from the centre of each segment, place these singly on the surface of the callus culture medium and incubate. Note the tissues of the stem from which the callus arises.

Further Reading

BURGESS, J. (1974). Towards Novel Plants. *New Scientist* **64**, 242–7.

BUTCHER, D. N. and STREET, H. E. (1964). Excised Root Culture. *The Botanical Review* **30**, 513–86.

COCKING, E. C. (1972). Plant Cell Protoplasts—Isolation and Development. *Annual Review of Plant Physiology* **23**, 29–50.

INGRAM, D. S. (1976). Growth of Biotrophic Parasites in Tissue Culture. In: *Encyclopedia of Plant Phsyiology: Physiological Plant Pathology*, edited by HEITEFUSS, R. and WILLIAMS, P. H. Springer, Berlin.

MURASHIGE, T. (1974). Plant Propagation through Tissue Cultures. *Annual Review of Plant Physiology* **25**, 135–66.

POWER, J. B. and COCKING, E. C. (1971). Fusion of Plant Protoplasts. *Science Progress*, Oxford **59**, 181–98.

SMITH, H. H. (1974). Model Systems for Somatic Cell Plant Genetics. *BioScience* **24**, 269–76.

STEWARD, F. C., MAPES, M. O., KENT, A. E. and HOLSTEN, R. D. (1964). Growth and Development of Cultured Plant Cells. *Science* **143**, 20–7.

STREET, H. E. (1969). Growth in Organised and Unorganised Systems. Knowledge Gained by Culture of Organs and Tissue Explants. In: *Plant Physiology—A treatise*. Edited by F. C. Steward. Volume VB. Academic Press, New York.

STREET, H. E. Editor (1973). *Plant Tissue and Cell Culture*. Blackwell Scientific Publications, Oxford.

SUNDERLAND, N. (1970). Pollen Plants and their Significance. *New Scientist* **47**, 142–4.

SUNDERLAND, N. (1971). Anther Culture: a Progress Report. *Science Progress* **59**, 527–49.

WHITE, P. R. (1963). *The Cultivation of Animal and Plant Cells*. 2nd Edition. Ronald Press, New York.

WILLMER, E. N. Editor (1966). *Cells and Tissues in Culture*. Volume 3. Chapters 8, 9 and 10. Academic Press, London and New York.

DUE